Christ Church
R002052

D1555753

The chemistry of enamines

Cambridge Chemistry Texts

The chemistry of enamines

S. F. DYKE
Reader in Organic Chemistry
University of Bath

CAMBRIDGE
at the University Press, 1973

Published by the Syndics of the Cambridge University Press
Bentley House, 200 Euston Road, London NW1 2DB
American Branch: 32 East 57th Street, New York, N.Y. 10022

Library of Congress Catalogue Card Number: 72–78893

ISBNs: 0 521 08676 0 hard covers
0 521 09731 2 paperback

Made and printed by offset in Great Britain by
William Clowes & Sons, Limited, London, Beccles and Colchester

Contents

Preface *page* vii
1 Preparation and some properties 1
1.1 *Preparation* 1
1.1.1 *From aldehydes and ketones* 1
1.1.2 *From ketals* 5
1.1.3 *From imines* 5
1.1.4 *From iminium salts* 6
1.2 *Some properties of enamines* 7
1.2.1 *Protonation and basicity* 7
1.2.2 *Ultraviolet spectra* 9
1.2.3 *Infrared spectra* 10
1.2.4 *Nuclear magnetic resonance spectra* 10
1.3 *Structures of isomeric enamines* 10
2 Reactions 14
2.1 *Electrophilic reagents* 14
2.1.1 *Alkyl halides* 14
2.1.2 *Acyl halides* 22
2.1.3 *Electrophilic olefines* 31
2.1.4 *Miscellaneous electrophilic reagents* 41
2.2 *Cycloaddition* 46
2.2.1 *Reaction with acetylenes* 47
2.2.2 *The Diels–Alder reaction* 49
2.2.3 *1,3-Dipolar additions* 50
2.2.4 *Reaction with carbenes* 50
2.2.5 *Other cycloadditions* 51
2.3 *Nucleophilic reagents* 51
2.3.1 *Organometallic compounds* 52
2.3.2 *Hydride ions* 52
2.3.3 *Reaction with diazoalkanes* 52
2.3.4 *Cyanide ion* 54
3 Heterocyclic enamines 55

3.1 *Definition* 55
3.2 *Preparation* 55
3.2.1 *Condensation of a carbonyl compound with a secondary amine* 55
3.2.2 *Oxidation of tertiary amines with mercuric salts* 56
3.2.3 *Partial reduction of aromatic nitrogen heterocycles* 61
3.2.4 *From lactams* 65
3.2.5 *From nitriles* 66
3.2.6 *By Claisen condensation* 66
3.3 *Structure* 67
3.3.1 *Secondary amines* 67
3.3.2 *Azabicycloalkane enamines* 68
3.4 *Reactions* 69
3.4.1 *Electrophilic reagents* 69
3.4.2 *Nucleophilic reagents* 73
3.4.3 *Cycloaddition reactions* 76
3.5 *Heterocyclic enamines in synthesis* 78
3.5.1 *Pyrrolines* 78
3.5.2 *Pyridine derivatives* 80
3.5.3 *Isoquinoline derivatives* 86
Bibliography 90
Index 91

Preface

The term 'enamine' was first introduced in 1927 to emphasise the structural similarity between the α,β-unsaturated amine system and the α,β-unsaturated alcohol moiety present in enols. Isolated reports concerning the reactions of enamines date back to the early nineteen hundreds; indeed in 1916 Robinson correctly interpreted the course of the reaction between an alkyl halide and ethyl β-aminocrotonate (R. Robinson, *J. Chem. Soc.*, 1916, **109**, 1038; see also E. E. P. Hamilton and R. Robinson, *ibid.*, p. 1029). However, it was not until 1954, when Stork and his associates described alkylation and acylation reactions, and demonstrated the ease of preparation of a number of enamines, that general interest was aroused. Since then a considerable amount of work has been reported on a wide variety of enamines, expanding considerably the scope of the original observations. This interest still continues as new and synthetically useful reactions of vinylamines are reported.

In this book which, like other volumes in the series, is aimed at the senior undergraduate and immediate postgraduate worker, an attempt is made to illustrate the basic principles of enamine chemistry. It is impossible to present a comprehensive survey in a volume of reasonable size, therefore emphasis has been placed upon those reactions that seem to offer useful synthetic procedures. It has also been necessary to exclude virtually all descriptions of dienamines, α,β-acetylenic amines (ynamines) and vinylogous amides. References to the original literature have been kept to a minimum; fortunately several recent reviews of all or part of the subject matter are available, and reference to these has been made for further reading.

It is my pleasure to thank Dr R. C. Brown for critically reading the manuscript, and for helping with the proof-reading.

S. F. DYKE

1 Preparation and some properties

1.1. Preparation

1.1.1. From aldehydes and ketones. The most versatile, and most often used method of formation of enamines involves the condensation between an aldehyde or ketone and a secondary amine:

$$\mathrm{>\overset{H}{C}{-}\overset{|}{C}{=}O + HN<} \rightleftharpoons \mathrm{>C{=}\overset{|}{C}{-}N<} + H_2O$$

Primary amines, of course, also react with carbonyl compounds to form imines:

$$\mathrm{>\overset{H}{C}{-}\overset{|}{C}{=}O + H_2N{-}} \rightleftharpoons \mathrm{>\overset{H}{C}{-}\overset{|}{C}{=}N{-}} + H_2O$$

and whereas tautomerism is possible with the formation of the enamine:

$$\mathrm{>\overset{H}{C}{-}\overset{|}{C}{=}N{-}} \rightleftharpoons \mathrm{>C{=}\overset{|}{C}{-}N\overset{H}{<}}$$

the equilibrium is almost always in favour of the imine structure. This book will be concerned almost entirely with tertiary enamines, derived from secondary amines.

The commonest procedure for enamine formation, which was first used by Herr and Heyl in 1952, then exploited by Stork *et al.*, involved heating under reflux an equimolecular mixture of the carbonyl compound and the amine in a solvent such as anhydrous benzene, toluene or xylene, with azeotropic removal of the water. Usually a Dean and Stark trap is used, but a number of modifications have been made (e.g. other drying agents, molecular sieves, etc.). Slow reactions can be catalysed by acids, usually *p*-toluenesulphonic acid (*p*-TSA). However, in the original preparation of enamines, described in 1936 by Mannich and Davidson, anhydrous potassium carbonate was used as condensing agent. The mechanism of the reaction is usually expressed as shown in scheme 1.1. In the acid-catalysed reaction, presumably the *N*-hemiacetal (I) is protonated to (III). The actual rate of enamine formation depends in a complex way upon several factors:

(i) the basicity of the amine;
(ii) the degree of steric hindrance in either the amine or the ketone, which affects the rate of formation of (I);
(iii) the rate of loss of the hydroxyl group from (I) or (III); and
(iv) the rate of loss of a proton from (II).

The enamine, once formed may be isolated, for example by distillation, or may be used *in situ* for subsequent reactions.

Scheme 1.1

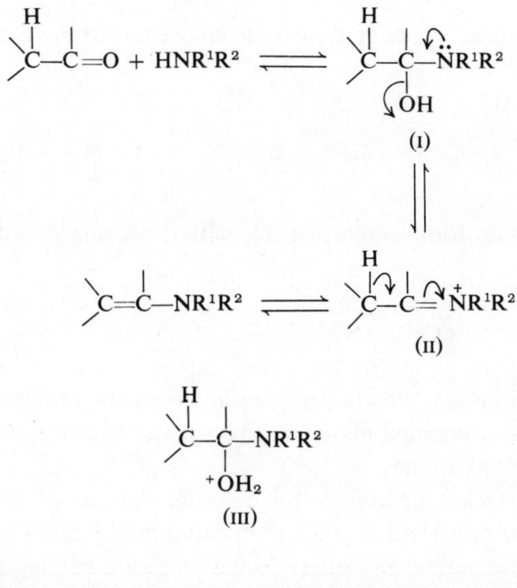

$$>C(H)-C=O + HNR^1R^2 \rightleftharpoons >C(H)-C(OH)-\ddot{N}R^1R^2 \quad (I)$$

$$(I) \rightleftharpoons >C(H)-C=\overset{+}{N}R^1R^2 \quad (II)$$

$$>C=C-NR^1R^2 \rightleftharpoons (II)$$

$$>C(H)-C(\overset{+}{O}H_2)-NR^1R^2 \quad (III)$$

The secondary amine. The most commonly used compounds have been the cyclic amines pyrrolidine (IV), piperidine (V) and morpholine (VI), and open-chain amines such as *N*-methylaniline and diethylamine. The actual amine used does influence, to some extent, the reactivity of the enamine formed (see p. 11).

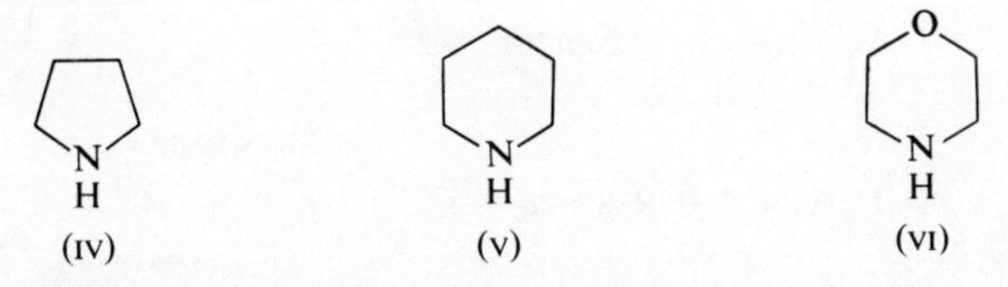

The carbonyl compound. This may be an aldehyde or ketone, and the latter may be acyclic or cyclic. Most studies of enamine chemistry have involved the use of cyclohexanone and cyclopentanone, but some open-chain ketones have also been used. Cyclohexanones without α-alkyl substituents react with pyrrolidine very readily (room temperature in methanolic solution), but α-alkylcyclohexanones, cycloheptanones and linear ketones react less readily. More hindered ketones, for example, 2,6-dialkylcyclohexanones, cannot be converted into enamines by azeotropic removal of water.

Side reactions have been observed when some methyl ketones react with secondary amines, e.g.

$$\mathrm{RCH_2\cdot CO\cdot Me} \xrightarrow[p\text{-TSA}]{\text{morpholine}} \mathrm{RCH_2{-}C(Me){=}CH\cdot CO\cdot CH_2R} + \mathrm{RCH_2{-}C(Me){=}CH{-}C(NC_4H_8O){=}CHR}$$

i.e. aldol condensation occurs to give the usual dimer, together with its enamine. The enamine of acetophenone is unstable and rapidly polymerises in the presence of a trace of acid. Other, less direct routes can be employed for the preparation of enamines from methyl ketones (see below).

The overall reaction scheme (scheme 1.2) for the formation of enamines of aldehydes is widely believed to involve the aminal (VII), but the evidence is conflicting. There is no doubt that aminals *are* produced, but it is not certain that they are the direct precursors in enamine formation.

Enamines derived from simple aldehydes are often unstable, being easily hydrolysed, oxidised or polymerised. The simplest possible ena-

Scheme 1.2

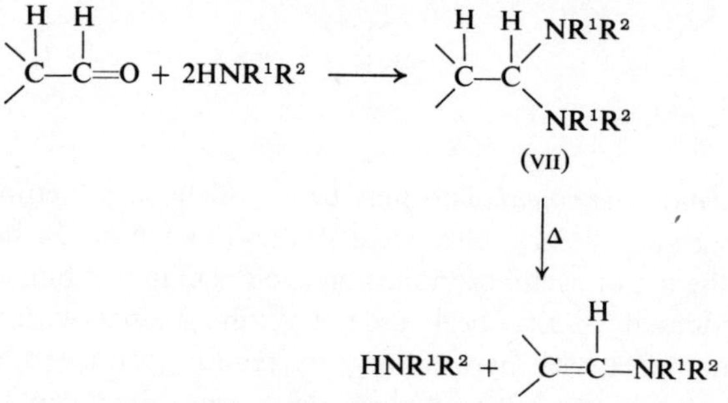

$$\geq C(H){-}C(H){=}O + 2HNR^1R^2 \longrightarrow \geq C(H){-}C(H)(NR^1R^2)_2 \quad (VII)$$

$$\xrightarrow{\Delta} HNR^1R^2 + \geq C{=}C(H){-}NR^1R^2$$

mine, that derived from acetaldehyde and dimethylamine, has been obtained by a modification of the method of Mannich and Davidson.

If a molecule contains both aldehydic and ketonic functions, e.g. (VIII), reaction with a secondary amine occurs at the former:

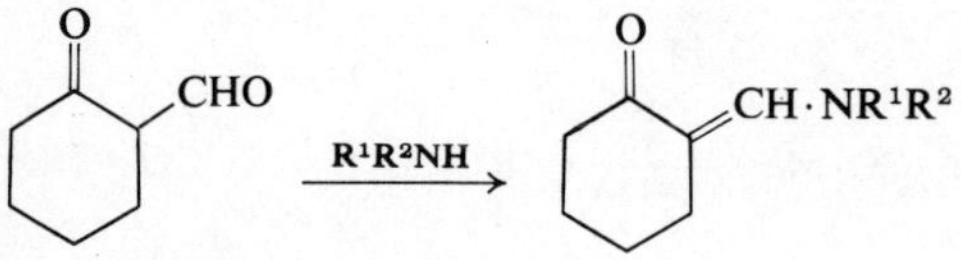

(VIII) $\xrightarrow{R^1R^2NH}$ CH·NR¹R²

Recently a very effective modification to the general procedure for preparing enamines was described. A stoichiometric mixture of titanium tetrachloride, the secondary amine and the carbonyl compound is left for several hours at room temperature. High yields of enamines are realised, especially from methyl ketones, hindered ketones and other carbonyl compounds. α-Tetralone (IX), which reacts with difficulty under normal conditions is converted into its enamine using titanium tetrachloride. Some other examples of enamine formation are included in scheme 1.3.

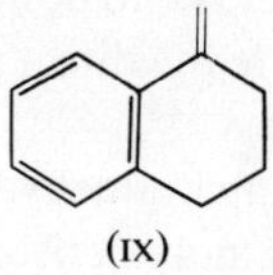

(IX)

Scheme 1.3

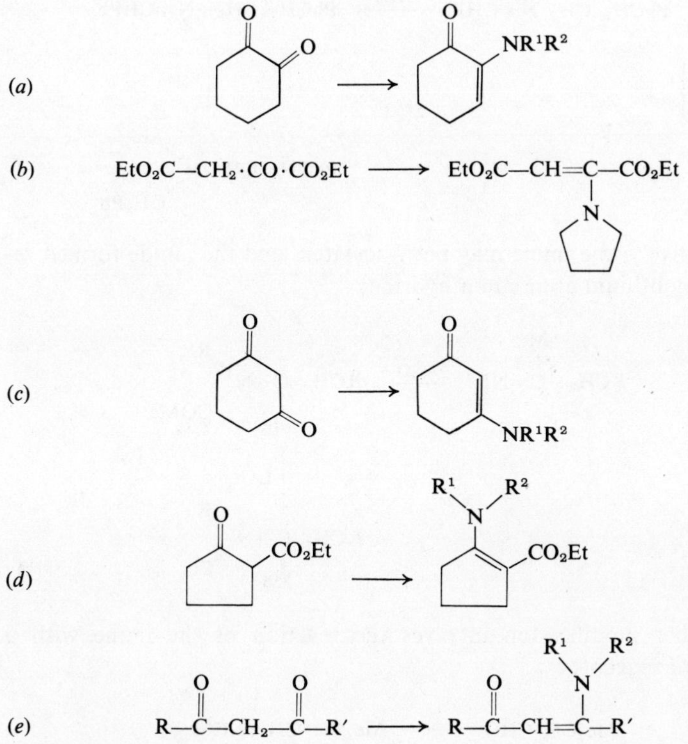

(*a*) (cyclohexane-1,2-dione) ⟶ 2-NR^1R^2-cyclohex-2-enone

(*b*) $EtO_2C—CH_2·CO·CO_2Et$ ⟶ $EtO_2C—CH{=}C(—N\text{(pyrrolidino)})—CO_2Et$

(*c*) (cyclohexane-1,3-dione) ⟶ 3-NR^1R^2-cyclohex-2-enone

(*d*) 2-CO_2Et-cyclopentanone ⟶ 1-(NR^1R^2)-2-CO_2Et-cyclopentene

(*e*) $R—C(=O)—CH_2—C(=O)—R'$ ⟶ $R—C(=O)—CH{=}C(NR^1R^2)—R'$

1.1.2. From ketals. Enamines have been obtained by heating secondary amines with ketals:

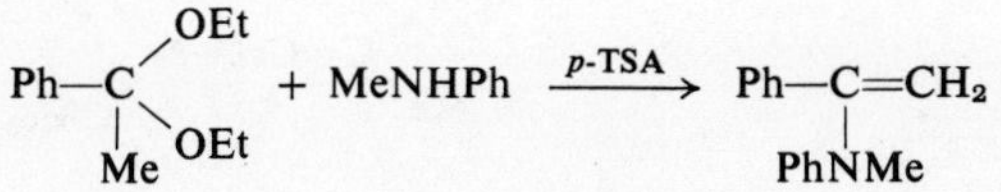

$$Ph—C(Me)(OEt)_2 + MeNHPh \xrightarrow{p\text{-TSA}} Ph—C(NMePh){=}CH_2$$

Although this method has not been studied very extensively, it has been used to obtain some enamines that are difficult to prepare by other methods.

1.1.3. From imines. The imine, prepared in the usual way from a carbonyl compound and a primary amine, can be converted into the tertiary enamine by a variety of methods. For example, by direct *N*-alkylation with an alkyl halide or a dialkyl sulphate, or by the use of the Meerwein reagent ($[Et_3O]^+BF_4^-$):

$$\mathrm{PhCH_2\cdot CH_2\cdot N{=}CHPh \xrightarrow{MeI} PhCH_2\cdot CH_2{-}\overset{\overset{\displaystyle Me}{|}}{\underset{\oplus}{N}}{=}CHPh}$$

$$\downarrow \text{Base}$$

$$\mathrm{PhCH{=}CH{-}N(Me)(CH_2Ph)}$$

Alternatively, the imine may be *N*-acylated, and the amide formed reduced by lithium aluminium hydride:

$$\mathrm{RCH_2{-}\overset{\overset{\displaystyle Me}{|}}{C}{=}NR' \xrightarrow{Ac_2O} RCH{=}\underset{\underset{\displaystyle Me}{|}}{C}{-}N(R')(COMe)}$$

$$\downarrow \text{LAH}$$

$$\mathrm{RCH{=}\underset{\underset{\displaystyle Me}{|}}{C}{-}N(R')(Et)}$$

Another modification involves the reaction of the imine with a Grignard reagent:

$$\mathrm{Me_2CH\cdot CHO \longrightarrow Me_2CH\cdot CH{=}NCMe_3}$$

$$\downarrow \text{EtMgBr/THF}$$

$$\mathrm{(Me)_2C{=}CH{-}N(MgBr)(CMe_3)}$$

1.1.4. From iminium salts. Since tertiary iminium salts (X) can be converted into the corresponding enamines (XI) by base, methods of preparation of the former are equivalent to the formation of the latter. A very

$$\underset{\text{(X)}}{\mathrm{>\overset{\overset{\displaystyle H}{|}}{C}{-}\overset{|}{C}{=}\overset{+}{N}(R^1)(R^2)}} \xrightarrow{\text{Base}} \underset{\text{(XI)}}{\mathrm{>C{=}\overset{|}{C}{-}NR^1R^2}}$$

simple and effective method, developed by Leonard and his associates, involves heating a ketone with the secondary amine perchlorate:

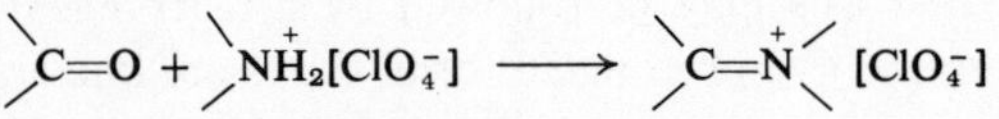

$$\rangle C{=}O + \rangle \overset{+}{N}H_2[ClO_4^-] \longrightarrow \rangle C{=}\overset{+}{N}\langle \; [ClO_4^-]$$

Three other useful methods exist for the preparation of enamines, but since they are of especial interest in the formation of heterocyclic enamines, discussion of them is deferred until Chapter 3. These methods involve: (*a*) oxidation of tertiary amines with mercuric acetate; (*b*) selective, partial reduction of aromatic nitrogen heterocycles; and (*c*) the reaction of lactams with Grignard reagents or with lithium aluminium hydride.

In the methods of formation of enamines described so far, a mixture of the two possible geometrical isomers (XII) and (XIII) is usually formed, often with the *trans* form predominating. It is, however, possible to

$$(R^1)(H)C{=}C(R^2)(NR_2) \quad \text{(XII)} \qquad (R^1)(H)C{=}C(NR^2)(R^2) \quad \text{(XIII)}$$

prepare the *cis*-isomer in a pure state by the stereospecific addition of an amine to an alkyne:

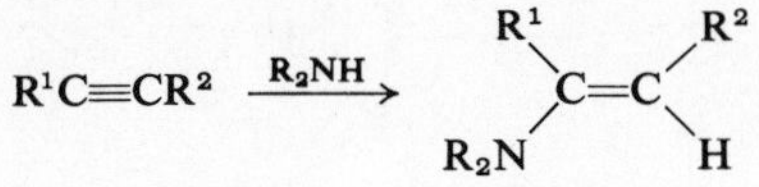

$$R^1C{\equiv}CR^2 \xrightarrow{R_2NH} (R^1)(R_2N)C{=}C(R^2)(H)$$

1.2. Some properties of enamines

1.2.1. Protonation and basicity. The structure of the enamine system may be regarded as a resonance hybrid to which the canonical forms (XIV) and (XV) are the important contributors. Hence, electrophilic

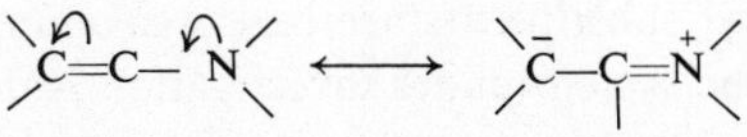

$$\rangle C{=}C{-}\ddot{N}\langle \;\text{(XIV)} \longleftrightarrow \rangle \bar{C}{-}C{=}\overset{+}{N}\langle \;\text{(XV)}$$

reagents, including protons, may attack the system at either the nitrogen atom, to give the ammonium salt (XVI), or, more importantly, at the carbon atom β to the nitrogen to yield an iminium salt (XVII). It is now known that the protonation of an enamine occurs initially at nitrogen, and that this may be followed by a rearrangement, that is usually rapid, to the *C*-β-protonated form. Cases are known, for example for derivatives of dehydroquinuclidine (XVIII), where *C*-protonation cannot take place for steric reasons.

One direct synthetic application of protonation reactions is the preparation of deuterated ketones, for example (XIX) is formed when the pyrrolidine enamine of 2-methylcyclohexanone is treated with deuteroacetic acid in deuterium oxide (CH_3CO_2D/D_2O).

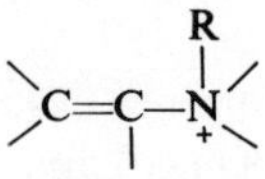

(XVI) 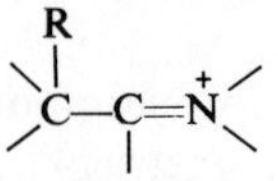 (XVII)

Of further interest is the fact that (XVII) is susceptible to attack by a nucleophilic reagent (Nu) to yield (XX). If R is a proton, the overall result is the addition of H—Nu to the carbon–carbon double bond of the enamine system.

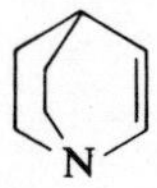

(XVIII) 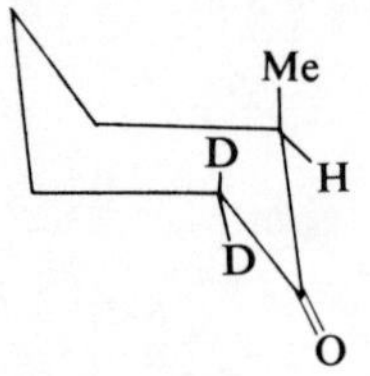 (XIX) 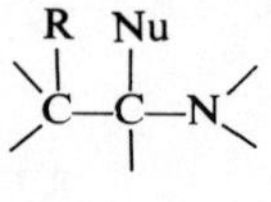 (XX)

Enamines possessing a tertiary nitrogen atom have often been stated to be more basic than the corresponding saturated amine, but recently this generalisation has been severely criticised. It seems that the presence or absence of an α-, or of a β-alkyl substituent in the enamine system will dictate the basicity. α-Alkyl substituents are base-strengthening whereas β-alkyl substituents are base-weakening. The question of base strength in the system under investigation is dependent upon the *position* of protonation. In some cases the *N*-protonated enamine formed initially does not rearrange to yield the *C*-protonated ion.

Enamines are usually easily hydrolysed by acids; the mechanism is often expressed (scheme 1.4) as a number of separate steps. The rate-determining step depends very much upon the pH of the medium and upon the structure of the original enamine.

Scheme 1.4

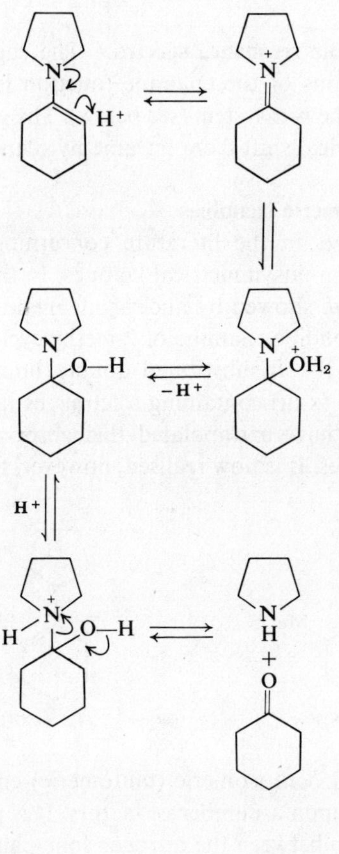

1.2.2. Ultraviolet spectra. Enamines give rise to a band at 220–240 nm (ϵ_{max} 5000–9000), depending upon the degree of *p*–π overlap between the nitrogen unshared electron pair and the carbon–carbon double bond.

There is little difference between the ultraviolet spectra of the *C*-β-protonated form, and the enamine, but if protonation occurs at nitrogen this ion would be expected to exhibit end absorption only, due to the isolated double bond.

1.2.3. Infrared spectra. The carbon–carbon double bond of the vinylamine function absorbs at about 1630–1660 cm^{-1} giving a medium-to-strong band. When the enamine is converted into its iminium salt by *C*-β-protonation, a very characteristic shift of 20–50 cm^{-1} to higher frequency occurs, with an increase in the intensity of the absorption.

1.2.4. Nuclear magnetic resonance spectra. The chemical shift of the vinyl proton or protons of the enamine function is indicative of the degree of overlap in the *p*–π system (see below). The greater the overlap, the lower is the chemical shift from tetramethylsilane (TMS).

1.3. Structures of isomeric enamines

Some confusion exists in the literature concerning the structure of enamines derived from unsymmetrical ketones. In their original definitive paper, Stork *et al.* showed by nuclear magnetic resonance (n.m.r.) studies that the pyrrolidine enamine of 2-methylcyclohexanone has the structure (XXI), with the trisubstituted double bond, rather than the more stable structure (XXII) containing a tetrasubstituted double bond. Subsequent workers have extrapolated this observation to other ketones and other amines. It is now realised, however, that in the majority

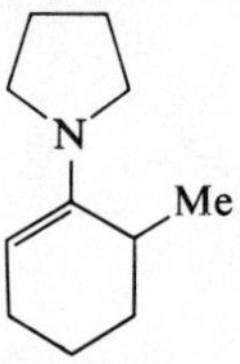

(XXI) 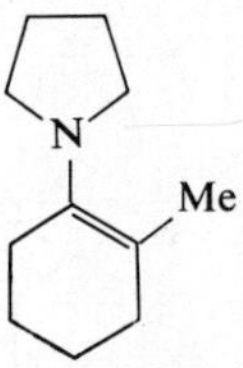 (XXII)

of cases a mixture of *both* isomeric (tautomeric) enamines is formed, the ratio depending upon a number of factors. It is postulated that the degree of *p*–π overlap between the nitrogen lone pair electrons and the double bond varies with the amine used to form the enamine of a given ketone, and is greater for the pyrrolidine enamine. This correlation is illustrated by the data collected into table 1.1. The first column of figures refers to the chemical shift of the olefinic proton in the cyclohexanone enamines formed from the amines indicated. The greater the degree of *p*–π overlap, the greater is the electron density at the β-carbon atom, and, consequently, the greater is the magnetic shielding of the vinylic proton. Thus, for 1-pyrrolidino-1-cyclohexene, the chem-

TABLE 1.1. *Nuclear magnetic resonance data for some enamines of cyclohexanone and 2-methylcyclohexanone*

Amine	δ for vinylic proton: 1-aminocyclohexene enamine	δ for vinylic proton: 2-methyl-1-aminocyclohexene enamine (Me)	Percentage of trisubstituted enamine
Pyrrolidine	4.17	4.17	90
Dimethylamine (Me–NH–Me)	4.36	4.43	60
Diethylamine (Et–NH–Et)	4.43	4.55	25
Morpholine	4.55	4.60	52
Piperidine	—	4.61	46

ical shift of the olefinic proton is at significantly higher field than that for the corresponding proton when any of the other amines is used to form the enamine. A similar correlation is observed (table 1.1) for the trisubstituted double bond isomer in the enamines derived from 2-methylcyclohexanone. The last column of the table indicates the proportion of the trisubstituted double bond isomer present at equilibrium for the various enamines of 2-methylcyclohexanone.

It is important to note that the greater the p–π overlap, the more reactive is the enamine to electrophilic reagents.

In an enamine with a high degree of *p*–π overlap, the system (XXIII) must be planar, or very nearly so. Steric hindrance may then occur, especially with the enamines formed from α-substituted ketones. In the

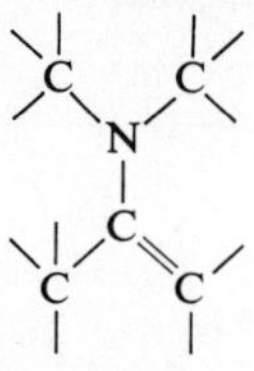

(XXIII)

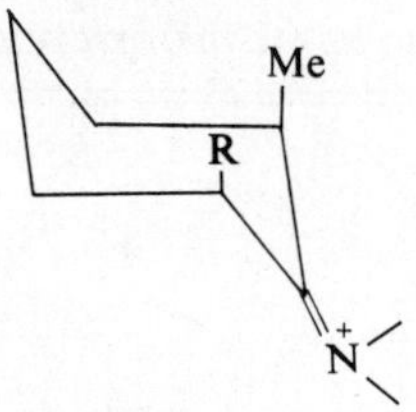

(XXIV)

particular case of the enamine from 2-methylcyclohexanone, it is known from other studies that the isomer with the trisubstituted double bond has an axial methyl group. Hence, in the pyrrolidine enamine, where the degree of p–π overlap is high, the trisubstituted enamine (XXI) is more stable than the tetrasubstituted isomer (XXII) because of the steric hindrance in the latter. Furthermore, the reaction of (XXI) to yield a 2,6-dialkylcyclohexanone derivative does not occur because of the severe 1,3-diaxial interaction that would be involved (XXIV).

Since the degree of p–π overlap is less for the enamines formed from other secondary amines, the steric hindrance is reduced, and the isomer containing the tetrasubstituted double bond becomes more stable, and therefore may predominate in the equilibrium.

Thus, the structure of a given enamine of an unsymmetrically substituted ketone is not unambiguously known – an important factor to bear in mind when syntheses involving enamines are being planned.

Some other, less obvious examples of enamine structures are illustrated in (XXV) and (XXVI). A mixture of isomeric enamines is formed

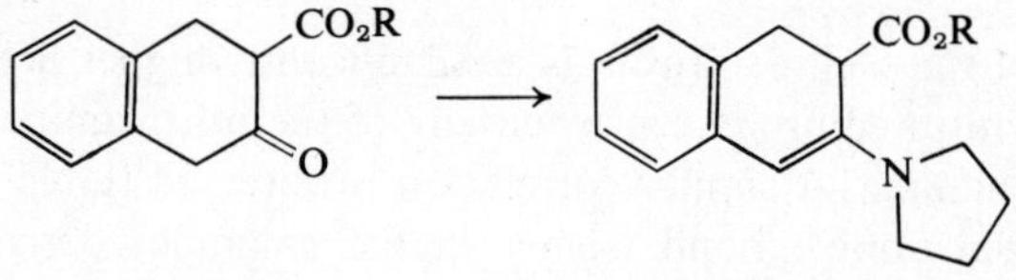

(XXV)

but

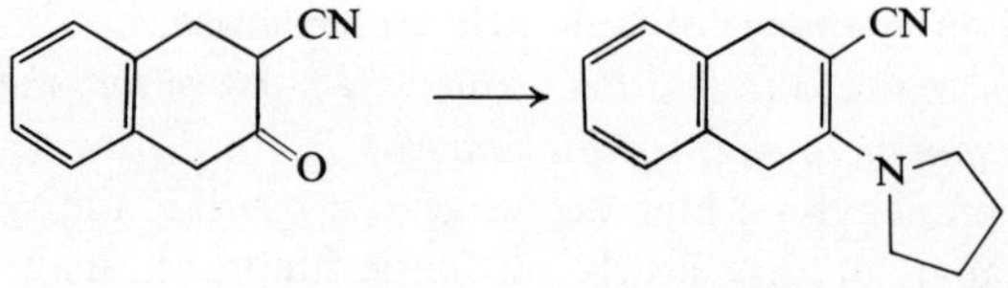

(XXVI)

from 3-methylcyclohexanones as well as from 2-methylcyclohexanone. Much less is known about the position of the double bond in enamines derived from 3-methylcyclopentanone. With aliphatic methyl ketones, a mixture is often obtained, depending upon the amine used to form the enamine. For example, the isomer (XXVII) is formed exclusively when methyl isopropyl ketone reacts with diethylamine but only to the extent of 35 per cent with morpholine; the other 65 per cent of the enamine mixture has structure (XXVIII).

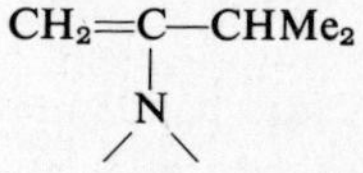

$$CH_2{=}C(N{<})\text{—}CHMe_2 \quad \text{(XXVII)} \qquad CH_3\text{—}C(N{<}){=}CMe_2 \quad \text{(XXVIII)}$$

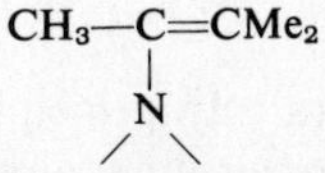

2 Reactions

2.1. Electrophilic reagents

2.1.1. Alkyl halides. In principle, an alkyl halide may attack an enamine at nitrogen, to give the salt (I), or at the carbon atom β to nitrogen, to yield the ion (II). It is the latter reaction that has synthetic usefulness.

$$\underset{\text{(I)}}{>C{=}C{-}\overset{+}{N}(R)<} \qquad \underset{\text{(II)}}{>C(R){-}C{=}\overset{+}{N}<}$$

Aldehyde enamines. When the enamines of acetaldehyde, or monosubstituted acetaldehydes react with alkyl halides, usually in boiling benzene solution, self-condensation products of the aldehyde are frequently the major products. If, however, a bulky amine such as di-t-butylamine is used to form the aldehyde enamine, carbon-alkylation occurs, often in good yields. Enamines derived from ββ-disubstituted

$$\underset{\text{(III)}}{Me_2C{=}CH{-}NMe_2} \xrightarrow{MeCH{=}CH\cdot CH_2Br} \underset{\text{(IV)}}{Me_2C{=}CH{-}\overset{+}{N}Me_2{-}CH_2{-}CH{=}CH(Me)} \longrightarrow \underset{\text{(V)}}{Me{-}CH(CH{=}CH_2){-}CMe_2{-}CH{=}\overset{+}{N}Me_2} \xrightarrow{\text{hydrolysis}} \underset{\text{(VI)}}{Me{-}CH(CH{=}CH_2){-}CMe_2{-}CHO}$$

aldehydes, such as $Me_2CH \cdot CHO$, are *N*-alkylated by simple alkyl halides.

When the more reactive allyl, propargyl or benzyl halides are used, *C*-alkylation predominates with most enamines derived from aldehydes (and also from open-chain ketones). This is because the initially formed *N*-alkyl compound can rearrange to the *C*-alkylated product. Such a rearrangement is an intramolecular process for the *N*-allyl or *N*-propargyl compounds (III) → (VI). This concept would explain the observation that when the enamine (VII) is alkylated with methyl *p*-toluenesulphonate, then hydrolysed, the product of the reaction is the aldehyde (IX); presumably the salt (VIII) is formed in the alkylation, and may then undergo rearrangement as shown.

$Me_2C{=}CH{-}N(Me){-}CH_2{-}CH{=}CH_2$ (VII) $\xrightarrow{p\text{-}MeC_6H_4SO_2OMe}$ $[Me_2C{=}CH{-}\overset{+}{N}Me_2{-}CH_2{-}CH{=}CH_2]$ (VIII)

(VIII) → $Me_2C(CH_2{-}CH{=}CH_2){-}CH{=}\overset{+}{N}Me_2$ $\xrightarrow{H_2O}$ $Me_2C(CH_2{-}CH{=}CH_2){-}CHO$ (IX)

When aldehyde enamines are alkylated with benzyl halides, the rearrangement from the *N*-benzylated product to the observed *C*-benzylated compound must be an intermolecular process. A mechanism for one such example is outlined in scheme 2.1, where a benzyl migration reaction must occur to account for the observation that methylation of the enamine (X), followed by hydrolysis, yields the aldehyde (XI).

Hence, the alkylation of aldehyde enamines, outlined above, makes available substituted aldehydes that would be very difficult, if not impossible, to obtain by the more traditional methods. However, very recently other new methods have been developed for the preparation of aldehydes (Carruthers, *Some Modern Methods of Organic Synthesis*, Cambridge Chemistry Texts, 1971).

Ketone enamines. It is with cyclic ketones, especially cyclohexanone and cyclopentanone, that the enamine alkylation reaction has been

Scheme 2.1

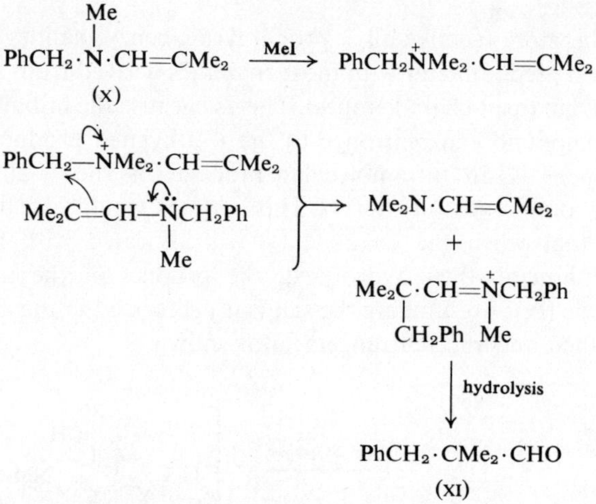

$$PhCH_2\cdot\underset{\text{(x)}}{\overset{Me}{\overset{|}{N}}}\cdot CH{=}CMe_2 \xrightarrow{MeI} PhCH_2\overset{+}{N}Me_2\cdot CH{=}CMe_2$$

$$\left.\begin{array}{l} PhCH_2{-}\overset{+}{N}Me_2\cdot CH{=}CMe_2 \\ Me_2C{=}CH{-}\ddot{N}CH_2Ph \\ \qquad\qquad\quad | \\ \qquad\qquad\quad Me \end{array}\right\} \longrightarrow \begin{array}{c} Me_2N\cdot CH{=}CMe_2 \\ + \\ Me_2C\cdot CH{=}\overset{+}{N}CH_2Ph \\ |\qquad\qquad | \\ CH_2Ph\quad Me \end{array}$$

$$\downarrow \text{hydrolysis}$$

$$\underset{\text{(xI)}}{PhCH_2\cdot CMe_2\cdot CHO}$$

most extensively studied, because of the importance of α-alkylated cyclic ketones for the synthesis of steroids and terpenoids. Some typical reactions of cyclohexanone enamines are summarised in scheme 2.2.

Scheme 2.2

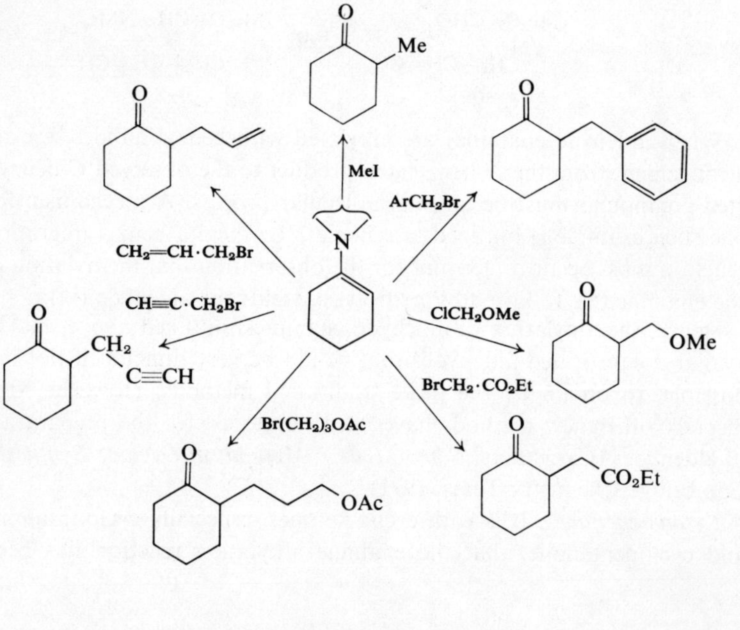

With the *pyrrolidine* enamine of cyclohexanone, alkylation with simple alkyl halides occurs predominantly at carbon; 2-methylcyclohexanone may be thus obtained in 80 per cent yield. However, for enamines derived from cycloheptanones, cyclooctanones or cyclononanones, *N*-alkylation predominates, as it does with the piperidine enamine of cyclohexanone. Polar solvents such as chloroform are better for *C*-alkylation than non-polar ones, such as dioxan. Alkylation is easier, and yields are higher for enamines prepared from more basic amines, although the ease of formation of the $>C=\overset{+}{N}<$ bond also influences the reaction. This latter point is exemplified by the pyrrolidine and piperidine enamines of cyclohexanone. Although pyrrolidine ($pK_a = 11.1$) and piperidine ($pK_a = 10.7$) have very similar basicities, the ~~minimum~~ iminium ion (XII) is formed more readily than (XIII).

(XII) (XIII)

Good yields of *C*-alkylated products are commonly obtained with the more reactive halides such as allyl, propargyl, and benzyl chlorides. There is evidence to indicate that *direct C*-alkylation occurs, rather than initial *N*-alkylation followed by rearrangement. For example, when 1-pyrrolidino-1-cyclohexene reacts with crotyl bromide, the products are (XIV) and (XV) in the ratio of 80:20.

1. $MeCH{=}CH \cdot CH_2Br$
2. hydrolysis

$CH_2 \cdot CH{=}CHMe$ (XIV)

+

Me
$CH \cdot CH{=}CH_2$ (XV)

An important feature of alkylation reactions of enamines is that usually the mono-alkylated ketone is obtained, either exclusively, or as the major product. Indeed, it was to solve the problem of the mono-alkylation of the substituted β-tetralone (XVI) that Stork examined the derived enamine, and thence proceeded to his fundamental study.

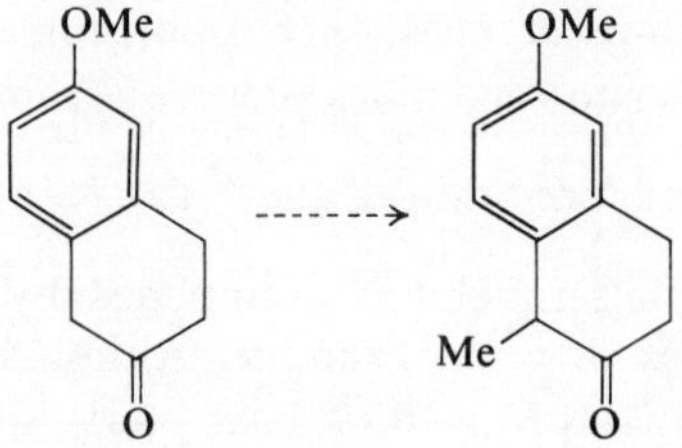

OMe OMe

Me

O O

(XVI)

In general, even when di-alkylation *does* occur, it is the symmetrical compound that is obtained:

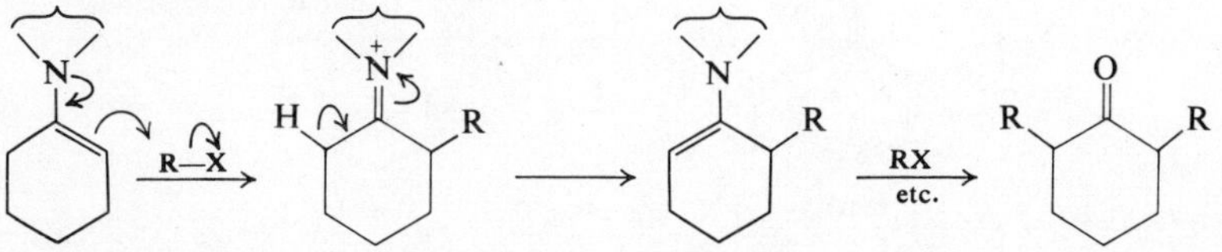

N H N R N R R O R

R—X RX etc.

The conditions required for the second alkylation step are usually much more severe than those needed for the initial alkylation, although in the presence of a strong base and an excess of alkylating agent, di-alkylation may predominate. Thus, two molar equivalents of allyl bromide will react with the pyrrolidine enamine of cyclohexanone in the presence of dicyclohexylethylamine. This amine is a strong base, but is sufficiently sterically hindered not to be alkylated by allyl bromide:

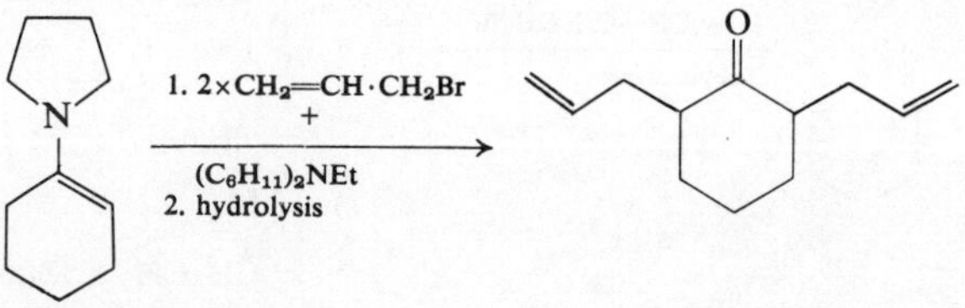

N

1. 2×CH_2═CH·CH_2Br
+
$(C_6H_{11})_2$NEt
2. hydrolysis

O

The di-alkylation reaction has been utilised for a simple synthesis of the novel bicyclic system (XVII):

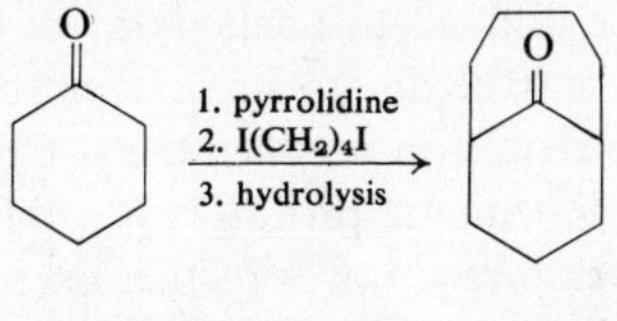

1. pyrrolidine
2. $I(CH_2)_4I$
3. hydrolysis

(15%)
(XVII)

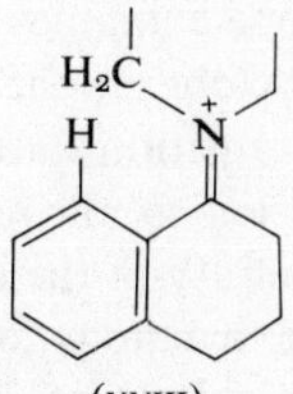

H_2C H $\overset{+}{N}$

(XVIII)

Scheme 2.3

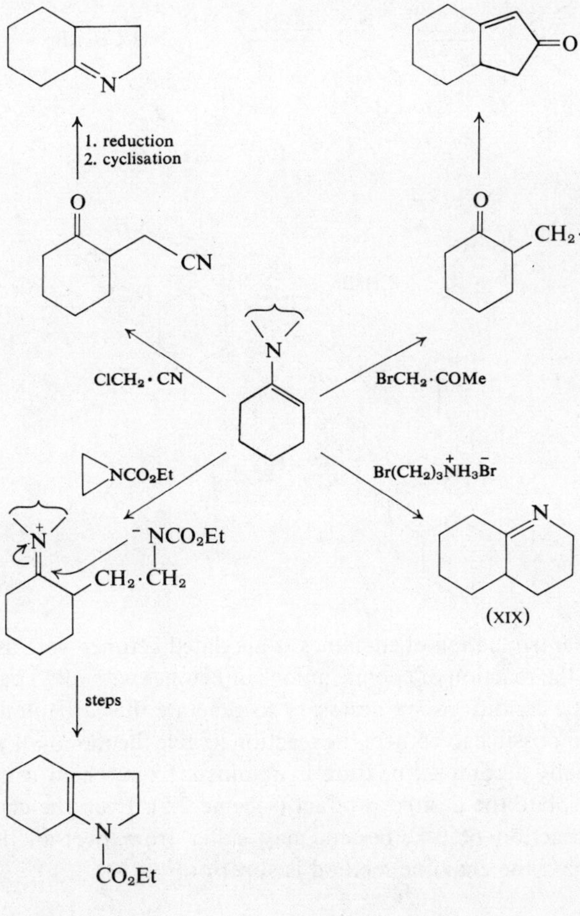

1. reduction
2. cyclisation

CN

$CH_2 \cdot COMe$

N

$ClCH_2 \cdot CN$

$BrCH_2 \cdot COMe$

NCO_2Et

$Br(CH_2)_3\overset{+}{N}H_3\overset{-}{Br}$

$\overset{+}{N}$

$^{-}NCO_2Et$

$CH_2 \cdot CH_2$

N

(XIX)

steps

N

CO_2Et

α-Tetralone enamines are much more difficult to *C*-alkylate than for example, cyclohexanone enamine; this may be due to steric overcrowding in the transition state (XVIII), where repulsion between the —CH_2— group adjacent to nitrogen and the *peri*-aromatic proton is possible.

The versatility of the enamine alkylation reaction is further exemplified by the reactions summarised in scheme 2.3. The formation of (XIX) is particularly noteworthy because the reaction involves an amine exchange reaction as shown in scheme 2.4.

Scheme 2.4

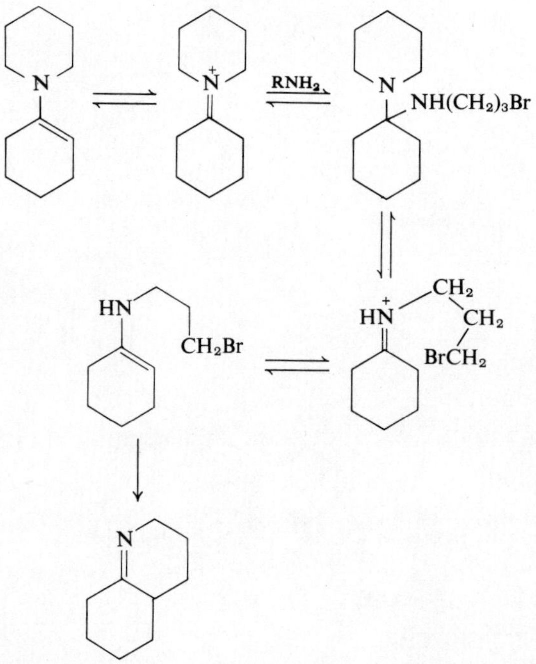

N N⁺ RNH_2 N $NH(CH_2)_3Br$

HN CH_2Br $H\overset{+}{N}$ CH_2 CH_2 $BrCH_2$

N

Before the introduction of enamines, α-alkylated ketones were usually prepared by the reaction of enolate anions of ketones with alkyl halides. Strongly basic conditions are necessary to generate the carbanion, and it is almost impossible to control the reaction to give the mono-alkylated ketone. Usually a complex mixture is obtained, from which it is very difficult to isolate the desired product (scheme 2.5). Even the classical alkylation reaction of β-ketoesters may suffer from over-alkylation, and once again the enamine method is superior.

Scheme 2.5

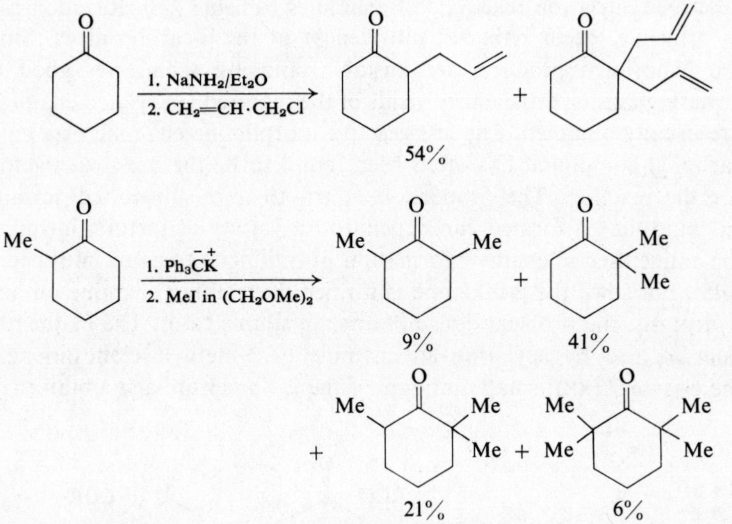

O

1. $NaNH_2/Et_2O$
2. $CH_2{=}CH{\cdot}CH_2Cl$

O + O

54%

Me O

1. $Ph_3\overset{-}{C}\overset{+}{K}$
2. MeI in $(CH_2OMe)_2$

Me O Me + O Me Me

9% 41%

+ Me O Me Me + Me Me O Me Me

21% 6%

Scheme 2.6

O N

PhCOCl

O $\overset{+}{N}$ O Ph

O N O Ph or O N O Ph

acid hydrolysis

O O Ph

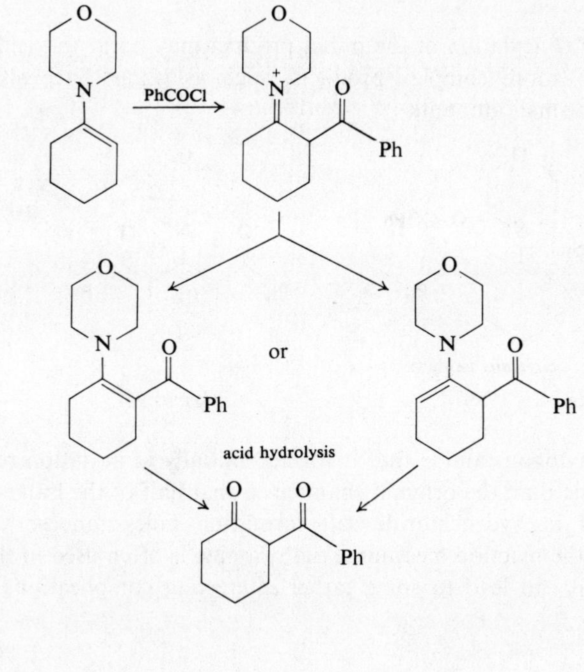

2.1.2. Acyl halides. In their original, definitive paper Stork *et al.* described acylation reactions of enamines (scheme 2.6). Reaction can, in principle, occur either at nitrogen or at the β-carbon atom. Since the *N*-acylated products are unstable, and are themselves good *C*-acylating agents, satisfactory yields of the required *C*-acylated enamines are usually obtained. The less reactive morpholine enamine of a given carbonyl compound has often been found to be the most satisfactory for the reaction. The proportion of tri- to tetra-substituted acylated enamine that is formed can depend on a variety of factors, including the amine used in enamine formation. It is difficult to enunciate general rules, but since the β-diketone is formed from either enamine on acid hydrolysis, the problem does not arise in simple cases. The major product formed by acylating an enamine of 3-methylcyclohexanone is the 6-isomer (XX); small amounts of the 2-isomer are also obtained.

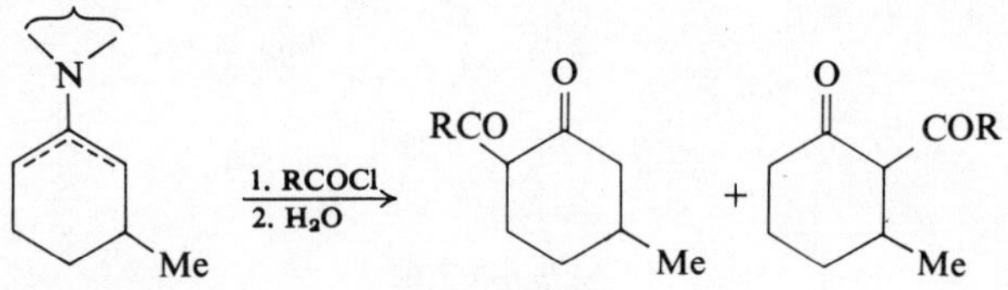

1. RCOCl, 2. H_2O

(XX) + (XXI)

Further *O*-acylation of the initial product may occur to yield (XXII), and other, more complex products, such as (XXIII) have also been isolated in small amounts.

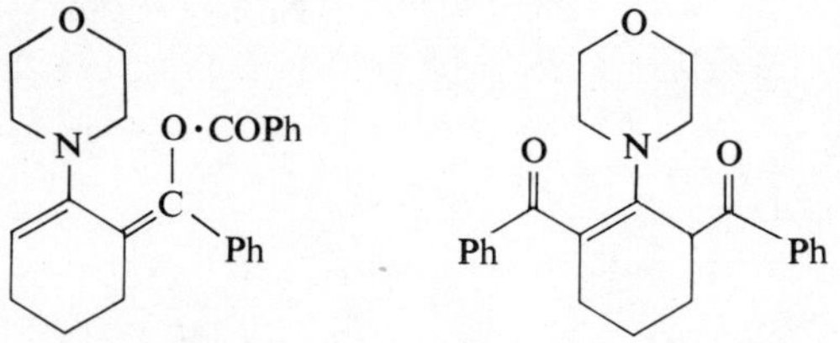

cis/trans isomers
(XXII)

(XXIII)

The vinylogous amide that is formed initially in acylation reactions is less basic than the original enamine so that half of the latter may be consumed in hydrochloride salt formation unless another base is added to the reaction medium. Triethylamine is often used in this context, but it can lead to some rather interesting complications. If the acid chloride possesses a α-hydrogen atom, a reaction occurs with triethylamine to yield a keten:

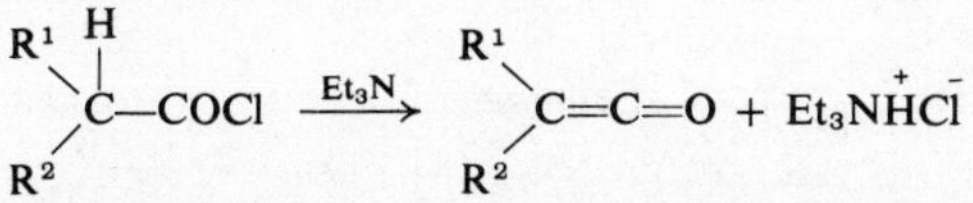

$$R^1R^2CH{-}COCl \xrightarrow{Et_3N} R^1R^2C{=}C{=}O + Et_3\overset{+}{N}H\overset{-}{Cl}$$

This keten may then undergo cycloaddition to the enamine to yield a cyclobutanone derivative:

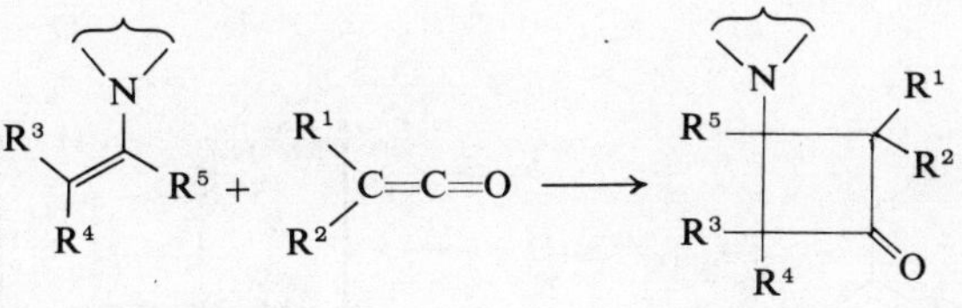

The fate of this product then depends upon the nature of the groups R^1–R^5. If R^1 or $R^2{=}H$, enolisation of the cyclobutanone can occur to give (XXIV), and thence, by ring-opening, the acylated enamine (XXV) is formed, which is *not* the product expected from direct acyla-

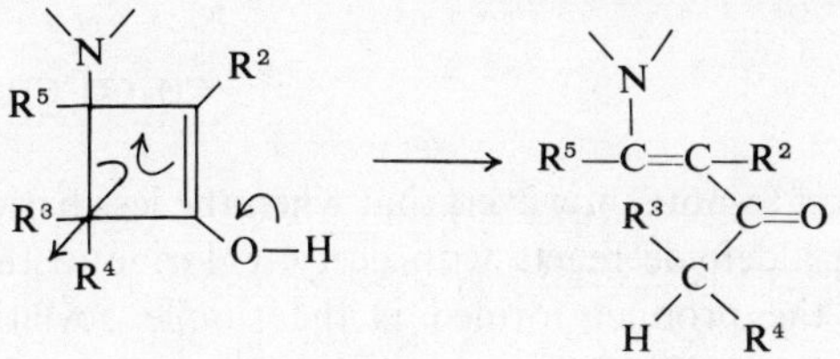

(XXIV) (XXV)

tion. Alternatively, if R^3 or $R^4{=}H$, the intermediate cyclobutanone can undergo ring-cleavage to (XXVII) *via* the enol (XXVI), to yield the acylated enamine expected from direct acylation. If R^1–R^4 are all

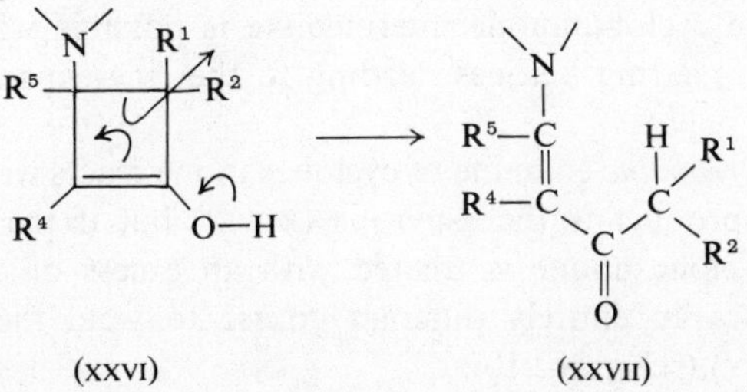

(XXVI) (XXVII)

alkyl groups, the cyclobutanone is thermally stable and can be isolated. Some illustrative examples, with aldehyde enamines ($R^5 = H$), are summarised in scheme 2.7.

Scheme 2.7

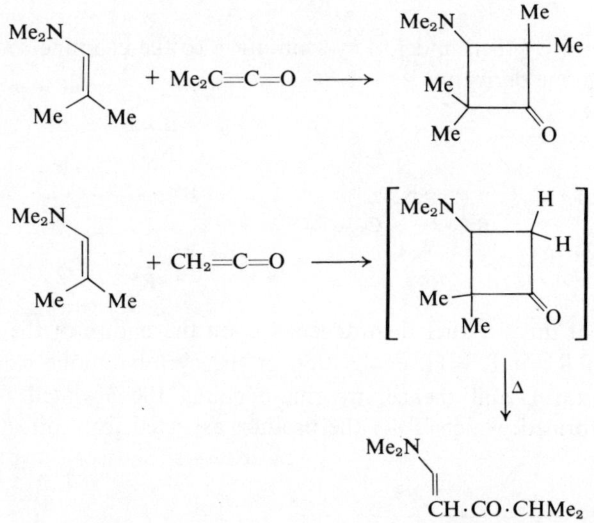

Me_2N–CH=C(Me)Me + $Me_2C{=}C{=}O$ ⟶ cyclobutanone (Me_2N, Me, Me, Me, Me, O)

Me_2N–CH=C(Me)Me + $CH_2{=}C{=}O$ ⟶ [cyclobutanone (Me_2N, H, H, Me, Me, O)] —Δ→ Me_2N–CH=CH·CO·$CHMe_2$

It is important to note, however, that when the less basic morpholine enamine of an aldehyde reacts with acetyl chloride in the *absence* of triethylamine, the product formed is the simple acylation product; keten, and consequent cyclobutanone formation are not involved.

There is some evidence to suggest that when the enamines of cyclic ketones, containing five, six or seven carbon atoms, are treated with acid chlorides that possess α-hydrogen atoms, in the presence of triethylamine, keten formation, cycloaddition and rearrangement occur (scheme 2.8), rather than direct acylation. An alternative mode of cleavage of the cyclobutanone intermediate is possible with the enamines from larger ring ketones, leading to the ring-expanded ketone (scheme 2.9).

When the *pyrrolidine* enamine of cyclohexanone reacts with an excess of keten, the product is the α-pyrone (XXVIII) but if the *morpholine* enamine of cyclohexanone is treated with an excess of diketen, the reaction follows an entirely different course to yield the chromone derivative (XXIX) (scheme 2.10).

Scheme 2.8

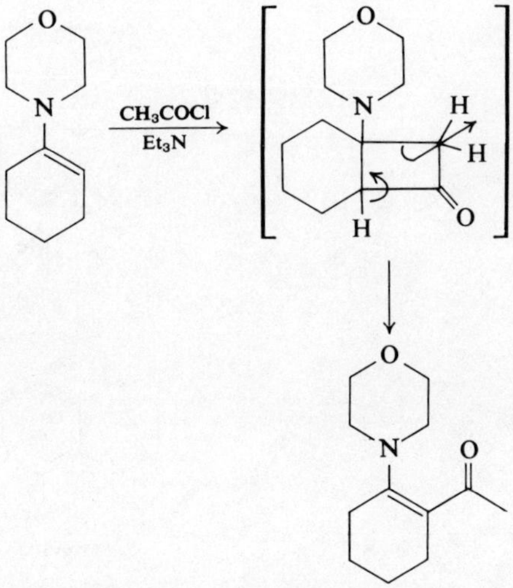

$\xrightarrow[Et_3N]{CH_3COCl}$

Scheme 2.9

$\xrightarrow[Et_3N]{CH_3COCl}$

$(CH_2)_{10}$ $(CH_2)_{10}$

$(CH_2)_{10}—CH_2$

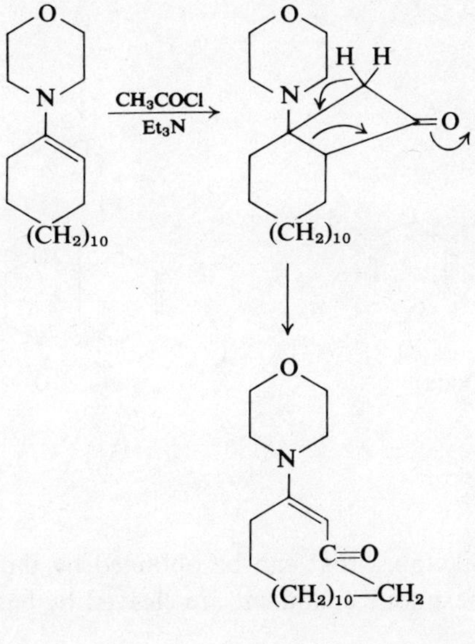

Scheme 2.10

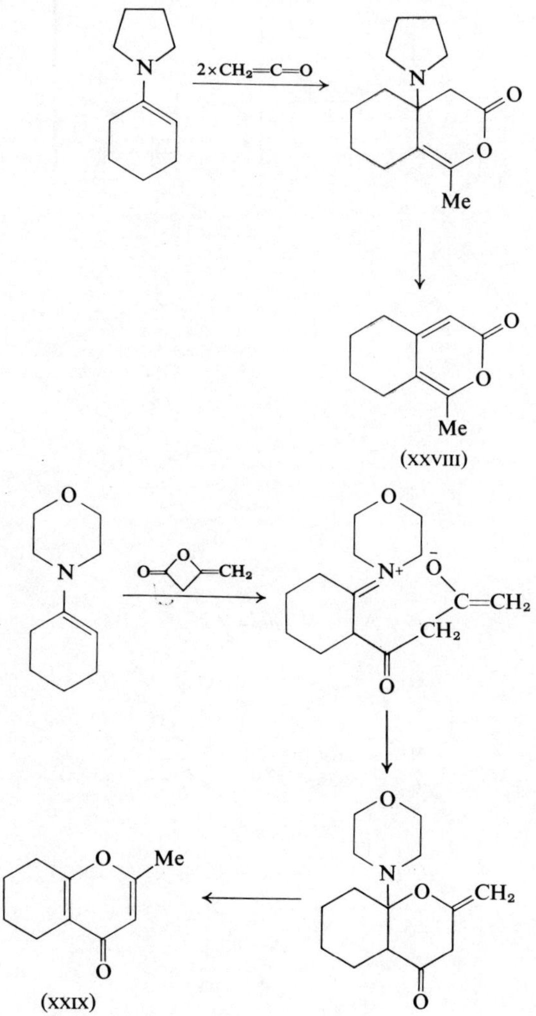

2×CH_2=C=O

Me

Me

(XXVIII)

O

N

CH_2

N+

C=CH_2

CH_2

O

Me

O

(XXIX)

CH_2

Since the β-diketones that can be obtained by the acylation of, for example, cyclohexanone enamines, are cleaved by base:

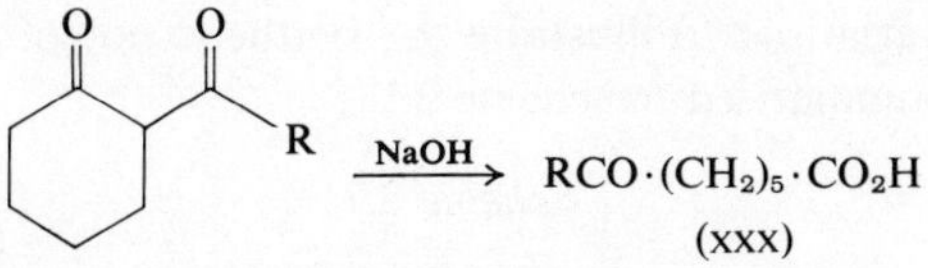
O
O
R
NaOH
RCO·(CH₂)₅·CO₂H

(xxx)

the sequence has been used in the synthesis of long-chain fatty acids. The keto-acid (xxx) can be easily reduced by the Clemmensen or Wolff-Kishner methods so that the overall scheme from cyclohexanone represents a method for ascending the acid series six carbon atoms at a time. By using long-chain dicarboxylic acid chlorides, the chain length can be extended very easily, provided that $n \geqslant 6$ in (xxxi).

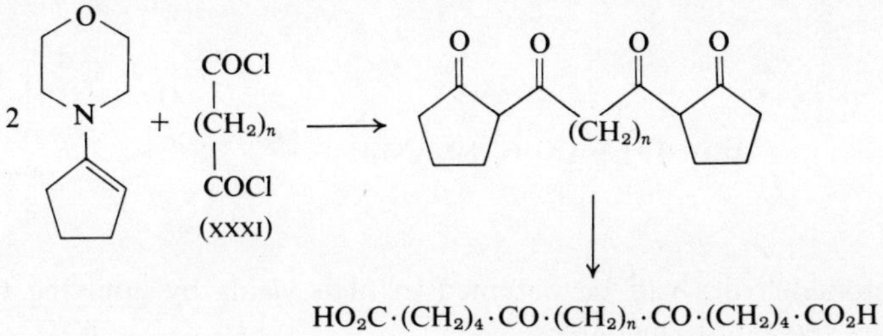
O
2
N
+
COCl
(CH₂)ₙ
COCl
(xxxi)
O
O
O
O
(CH₂)ₙ
HO₂C·(CH₂)₄·CO·(CH₂)ₙ·CO·(CH₂)₄·CO₂H

If $n \leqslant 5$ in (xxxi) the reaction takes a different course and leads, via the vinylogous acid anhydride (xxxii), to the keto-dibasic acid (xxxiii):

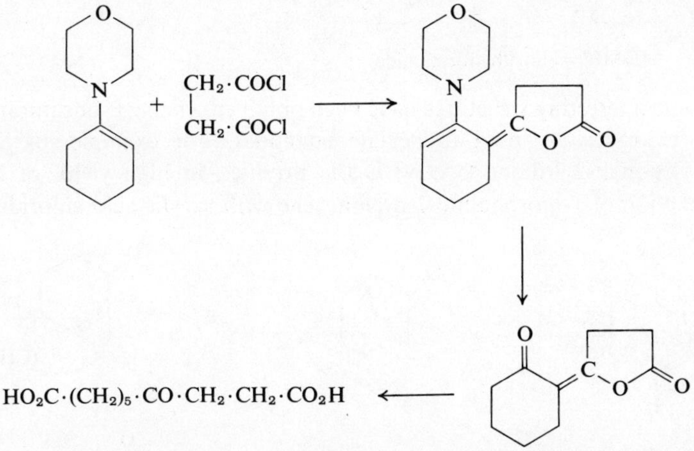
O
N
+
CH₂·COCl
CH₂·COCl
O
N
C
O
O
O
C
O
O
HO₂C·(CH₂)₅·CO·CH₂·CH₂·CO₂H

(xxxiii) (xxxii)

A further variation, to illustrate the synthetic potential of acylation reactions, is summarised in scheme 2.11.

Scheme 2.11

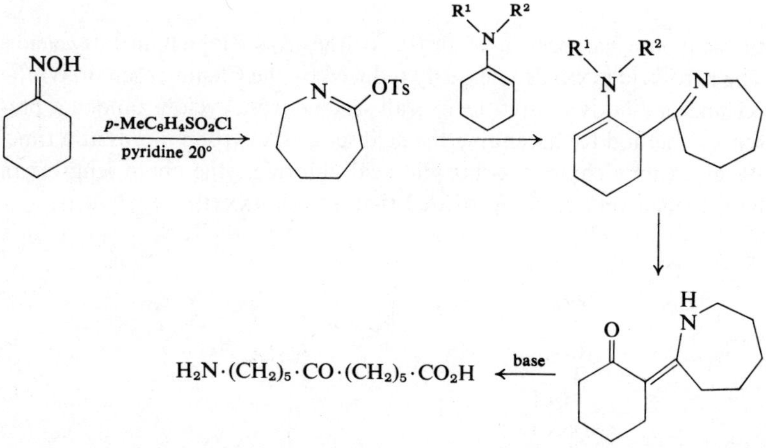

NOH

p-$MeC_6H_4SO_2Cl$
pyridine 20°

N OTs

R^1 R^2 N

R^1 R^2 N N

$H_2N \cdot (CH_2)_5 \cdot CO \cdot (CH_2)_5 \cdot CO_2H$ ← base

O H N

β-Keto-aldehydes can be obtained in high yields by applying the Vilsmeier reaction to enamines:

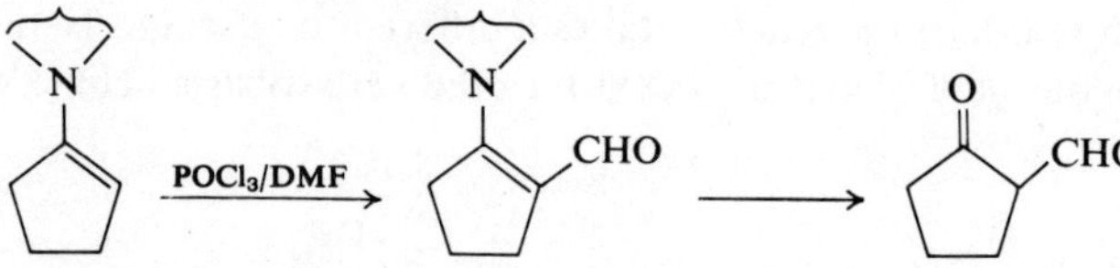

N

$POCl_3/DMF$

N CHO

O CHO

(DMF = dimethylformamide)

Some interesting structures have been obtained when α,β-unsaturated acid chlorides are used to acylate enamines. For example, bicyclo [3.3.1] nonan-2,9-dione (XXXIV) is the product, in high yield, of the interaction of 1-morpholino-1-cyclohexene with acrylic acid chloride:

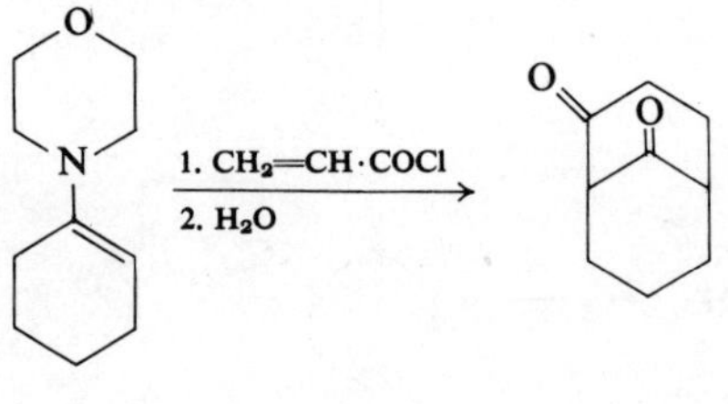

O N

1. $CH_2{=}CH \cdot COCl$
2. H_2O

O O

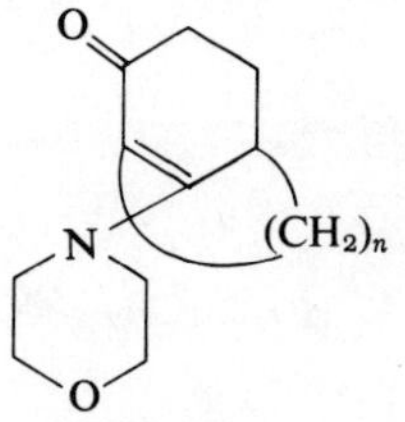

(XXXIV)

O N O $(CH_2)_n$

(XXXV)

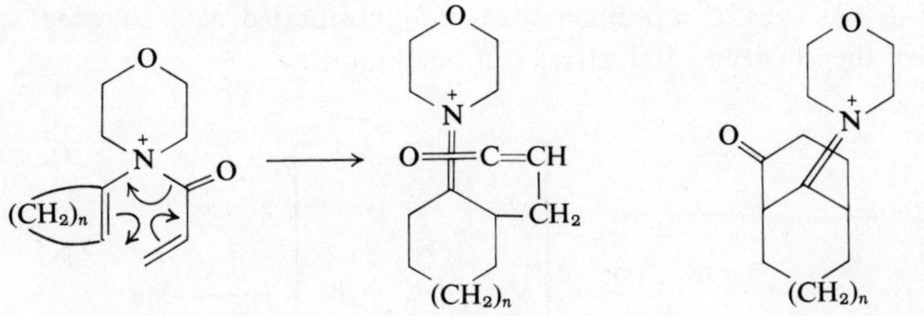

(XXXVI) (XXXVII) (XXXVIII)

If, however, the cyclic ketone used contains nine or ten carbon atoms, the product is of the type (XXXV). In these cases, *N*-acylation occurs initially to give (XXXVI), which then undergoes a [3,3]-sigmatropic rearrangement to the ion (XXXVII). The final product then depends upon the fate of this ion. When n is large, the one-carbon bridge can accommodate a double bond, leading to (XXXVIII), and the final product is (XXXV). When smaller ring ketones are used, for example $n = 4$, the diketone (XXXIV) is formed.

When the enamines of open-chain ketones, for example (XXXIX), react with α,β-unsaturated acid chlorides, the products are cyclohexa-

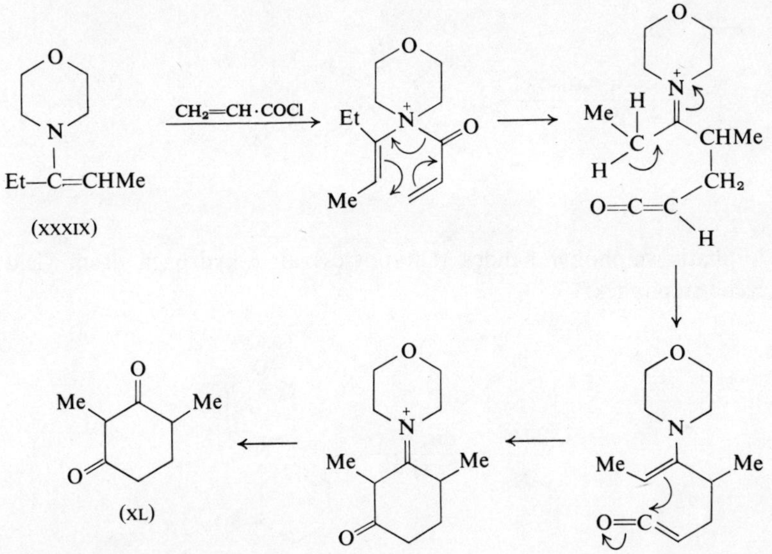

$CH_2{=}CH{\cdot}COCl$

(XXXIX) (XL)

1,3-diones (XL). If α,β-dihalogeno- α,β-unsaturated acid chlorides are used, then benzene derivatives can be obtained:

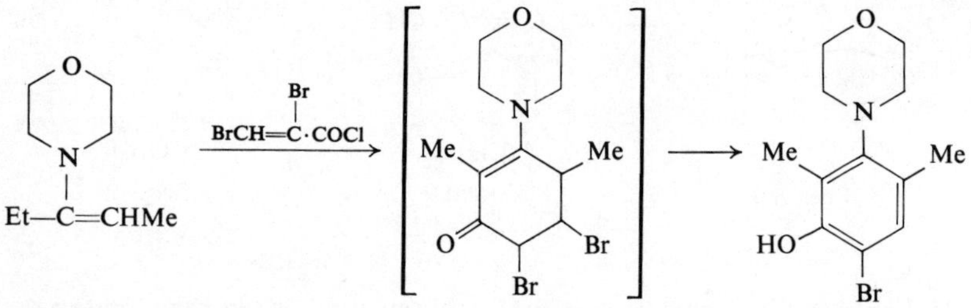

(XXXIX)

Aromatic sulphonyl halides also react readily with enamines to yield the expected sulphonated aldehyde or ketone:

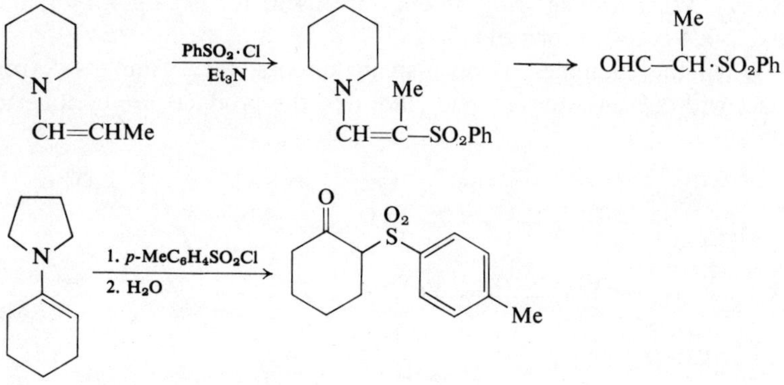

Aliphatic sulphonyl halides that possess an α-hydrogen atom yield cyclic sulphones:

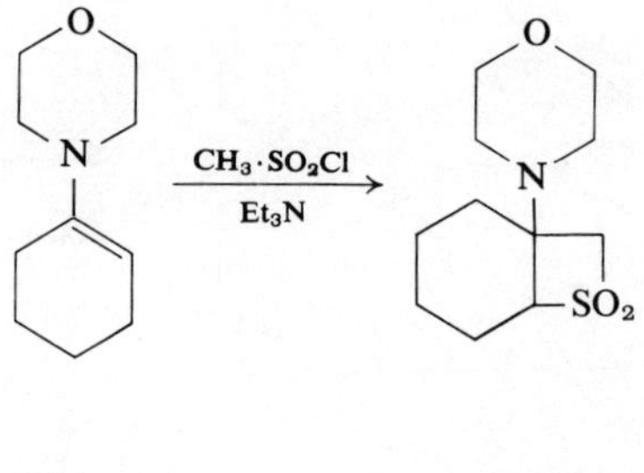

and it is postulated that a reaction between the sulphonyl halide and the triethylamine occurs initially to generate a sulphene:

$$CH_3 \cdot SO_2Cl + Et_3N \longrightarrow CH_2{=}\overset{\overset{\displaystyle O}{\|}}{S}{=}O + Et_3N\overset{+}{H}\overset{-}{Cl}$$

which is then involved in a cyclo-addition to the enamine.

2.1.3. Electrophilic olefines. Since enamines are nucleophilic reagents, they can react with electrophilic olefines (i.e. olefines in which an unsaturated, electron-attracting group is directly attached to the double bond) at the β-carbon atom (scheme 2.12). The undesired *N*-alkylation:

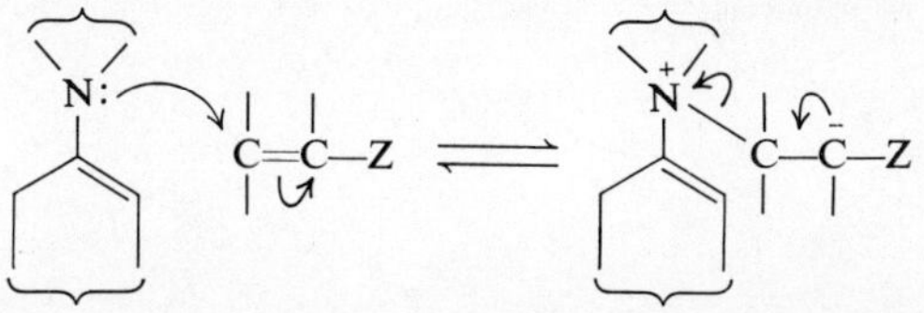

is reversible so that, usually, high yields of carbon-alkylation products may be obtained. The actual fate of the zwitterion (XLI) depends very

Scheme 2.12

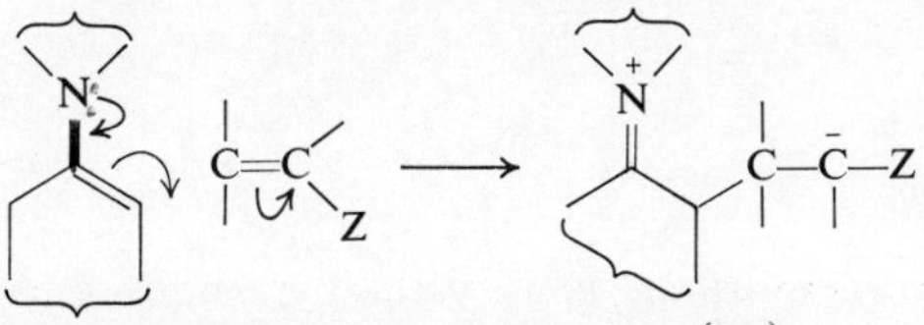

(XLI)

Z = CN, CO_2R, COR, CHO, NO_2, SO_2R, etc.

largely upon

(*a*) the type of enamine used – i.e. whether it is derived from an aldehyde or a ketone;

(*b*) the nature of the group Z attached to the double bond in the olefine; and

(*c*) the conditions of the reaction – temperature and solvent are especially important.

2.1.3a. Enamines of ketones

With α,β-unsaturated nitriles and esters. The zwitterion undergoes proton addition and proton loss to give the alkylated enamine; acid hydrolysis then yields the α-alkylated ketone (scheme 2.13). The con-

Scheme 2.13

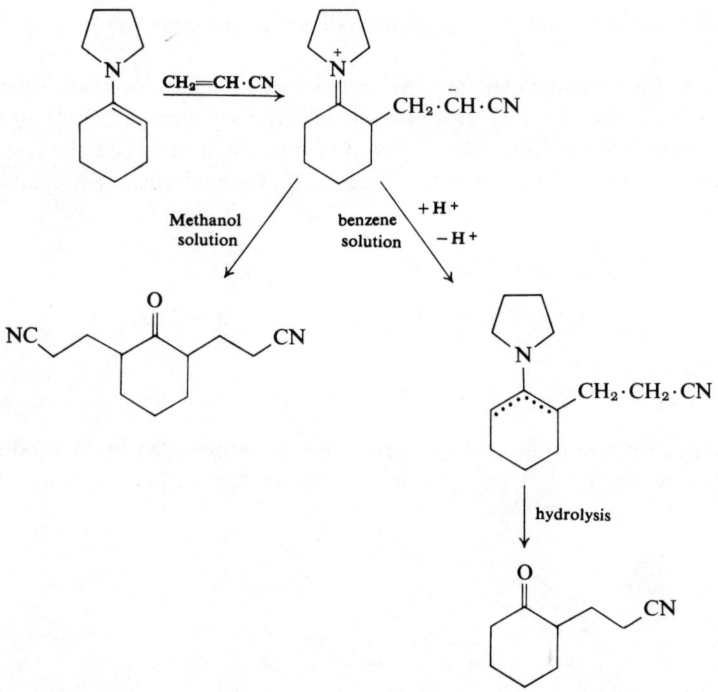

$CH_2{=}CH\cdot CN$

$CH_2\cdot \bar{C}H\cdot CN$

Methanol solution

benzene solution

$+H^+$

$-H^+$

NC

CN

$CH_2\cdot CH_2\cdot CN$

hydrolysis

CN

ditions used are important. Thus, if the pyrrolidine enamine of cyclohexanone is heated in benzene solution with acrylonitrile, the mono-alkylated product results but in ethanolic solution further reaction occurs to yield the symmetrical dialkylcyclohexanone.

If a straightforward Michael reaction between cyclohexanone and acrylonitrile is attempted, polyalkylation results.

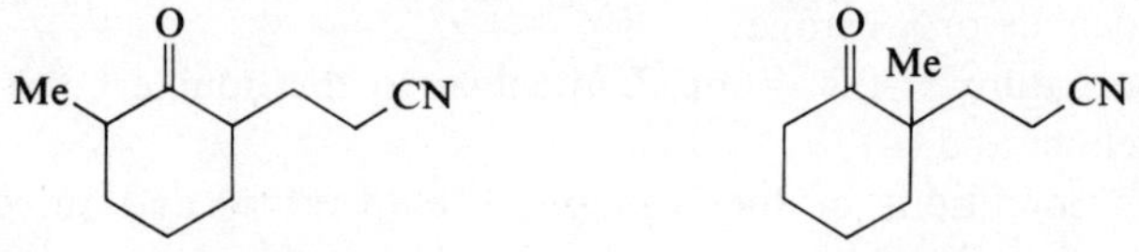

Me CN

(XLII)

Me CN

(XLIII)

When the enamine of 2-methylcyclohexanone is treated with acrylonitrile in ethanol solution, the product is (XLII), but in benzene solution equal amounts of (XLIII) are also produced.

Instead of the initially formed zwitterion (XLI) becoming neutralised

Scheme 2.14

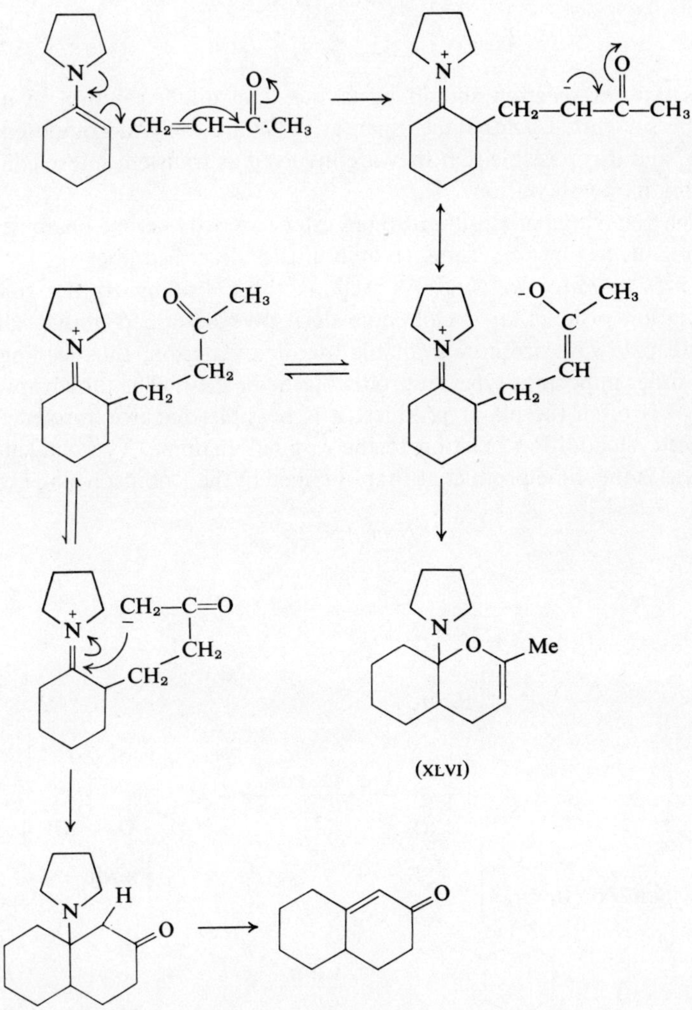

$CH_2{=}CH{-}C{\cdot}CH_3$ (=O)

$CH_2{-}CH{-}C{-}CH_3$ (=O)

O, CH_3, C, CH_2, CH_2

-O, CH_3, C, CH, CH_2

$CH_2{-}C{=}O$, CH_2, CH_2

N, O, Me

(XLVI)

N, H, O

O

(XLVII)

by proton loss and proton addition as indicated in scheme 2.13, an internal collapse is possible leading to a cyclobutane compound:

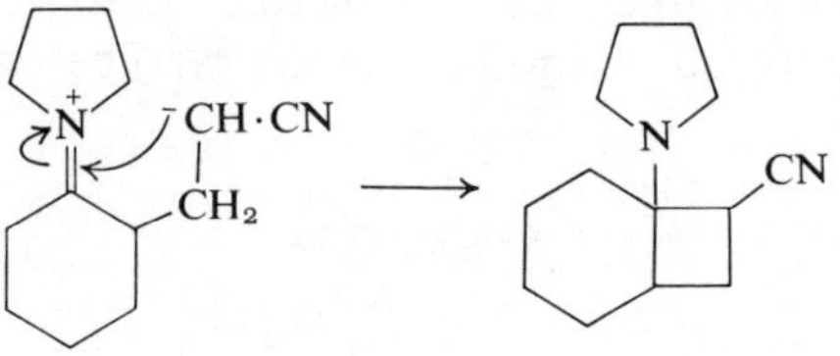

$\bar{C}H \cdot CN$ CH_2 → CN

(XLIV) (XLV)

This type of reaction should be favoured in dilute solution in non-protic solvents. Cyclobutanes such as (XLV) are easily decomposed by heat, and it is possible that they are involved as transient intermediates during mono-alkylation.

The behaviour of α,β-unsaturated esters towards ketone enamines is analogous to that of α,β-unsaturated nitriles described above.

α,β-*Unsaturated ketones.* As well as those leading to the simple alkylation product or cyclobutane derivative described above, alternative pathways are now available for the zwitterion, thus leading to two other important types of product (scheme 2.14). The dihydropyran (XLVI) is often the major product; it is possible that it is formed in a direct cycloaddition reaction to the original enamine. The octalenone (XLVII) is the same product as that obtained in the Robinson ring exten-

Scheme 2.15

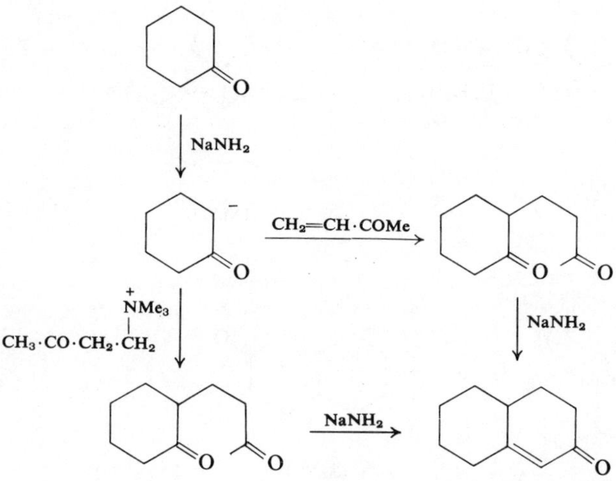

$NaNH_2$

$CH_2{=}CH \cdot COMe$

$CH_3 \cdot CO \cdot CH_2 \cdot CH_2\overset{+}{N}Me_3$

$NaNH_2$

$NaNH_2$

sion reaction, which is so important in a variety of schemes that have been evolved for the total synthesis of steroids and triterpenoids. Typically, a cyclohexanone is treated with either an α,β-unsaturated ketone such as methyl vinyl ketone, or its related Mannich base methiodide, in the presence of a strong base, such as sodamide (scheme 2.15). The diketone formed initially can be isolated, depending upon the conditions of the reaction.

The reaction sequence depicted in scheme 2.14 is somewhat more complex than that indicated, for it has been demonstrated that, prior to the final hydrolysis to form the ketonic products, the reaction mixture contains the enamines (XLVIII), (XLIX) and possibly (L) as well as

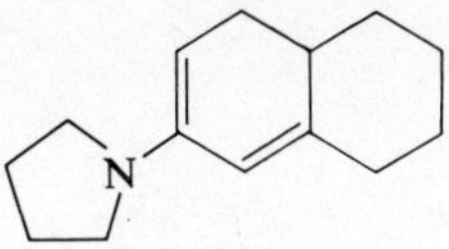

(XLVIII) 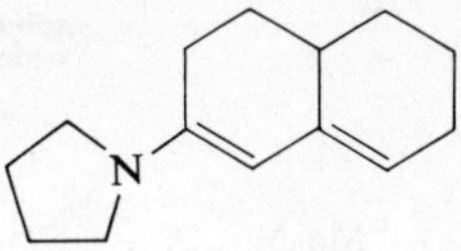 (XLIX)

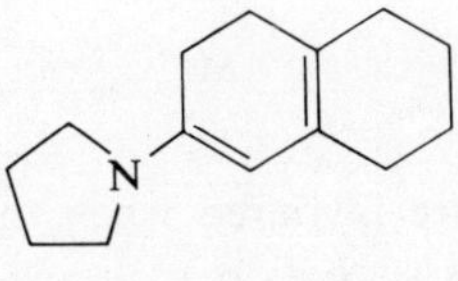

(L) 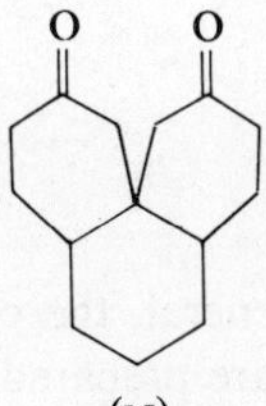 (LI)

the diketone (LI). The diketone is not formed from (XLVIII), (XLIX) or (L), and probably arises from the di-alkylation product as shown:

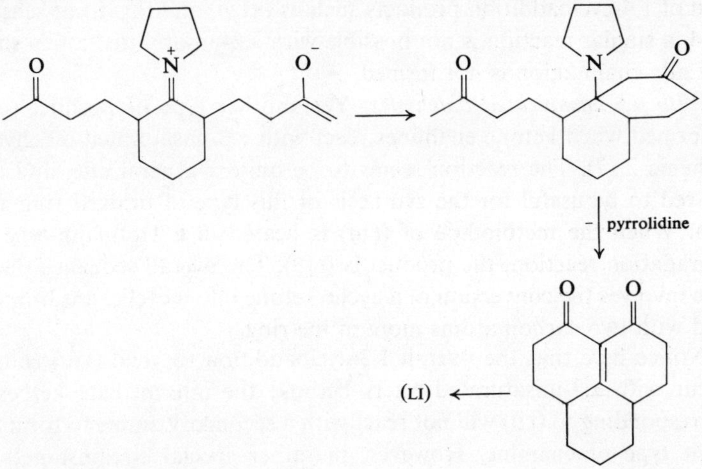

pyrrolidine

(LI)

When aliphatic ketone enamines react with methyl vinyl ketone, two products are possible, depending upon the temperature of the reaction (scheme 2.16).

Scheme 2.16

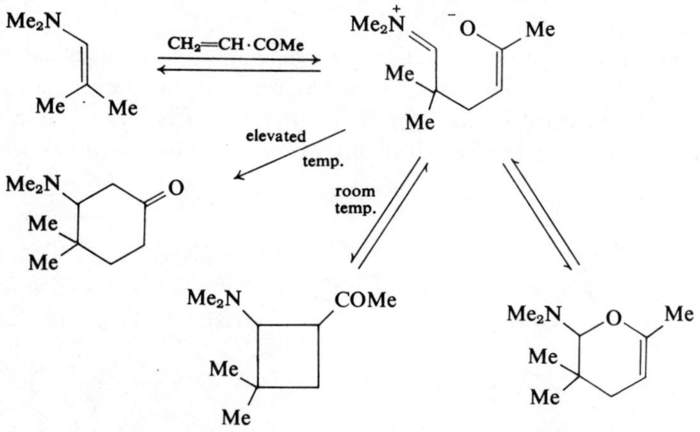

In general, the cycloaddition products are favoured when the enamines are prepared from the stronger bases such as pyrrolidine. Morpholine enamines give rise predominantly to the straightforward alkylation (or Michael addition) products.

Notice that, according to the mechanism postulated for the formation of 1,4-cycloaddition products such as (XLVI) and (XLVII) of scheme 2.14, a similar reaction is not possible with α,β-unsaturated esters since the side-chain anion is not formed.

With α,β-unsaturated aldehydes. Yet another type of product (LIII) is formed when ketone enamines react with α,β-unsaturated aldehydes (scheme 2.17). The reaction seems to be quite a general one, and has proved to be useful for the synthesis of this type of bridged ring system. When the methiodide of (LIII) is heated in a Hofmann-type of degradation reaction, the product is (LIV). The overall sequence therefore involves the conversion of a cyclic ketone into a cyclic, unsaturated acid with two carbon atoms more in the ring.

Notice here that the overall 1,3-cycloaddition to yield (LIV) cannot occur with α,β-unsaturated esters because the intermediate ketoester corresponding to (LII) will not react with a secondary amine to form the right type of enamine. However, in rather special circumstances, a

Scheme 2.17

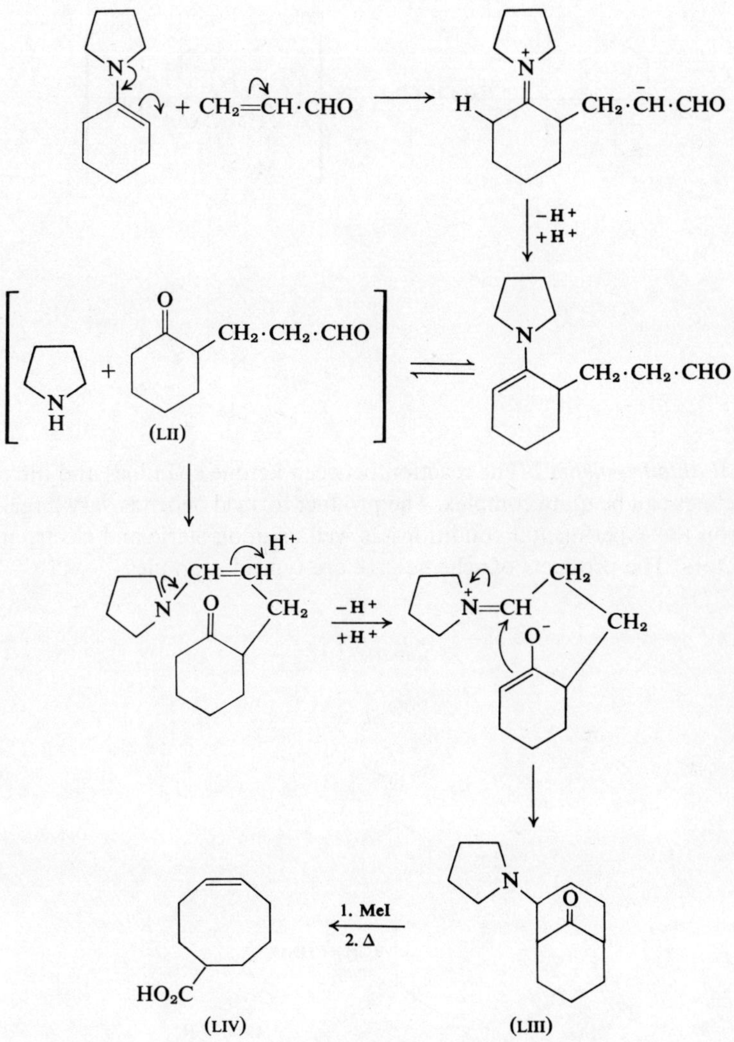

+ $CH_2{=}CH{\cdot}CHO$ ⟶ H $CH_2{\cdot}\bar{C}H{\cdot}CHO$

$-H^+$
$+H^+$

$CH_2{\cdot}CH_2{\cdot}CHO$ ⇌ $CH_2{\cdot}CH_2{\cdot}CHO$

N
H
+
(LII)

H^+
CH=CH
N O CH_2 $\underset{+H^+}{\xrightarrow{-H^+}}$ $\overset{+}{N}$=CH CH_2 CH_2 O^-

1. MeI
2. Δ

HO_2C

(LIV) (LIII)

different type of 1,3-cycloaddition involving α,β-unsaturated esters can be observed:

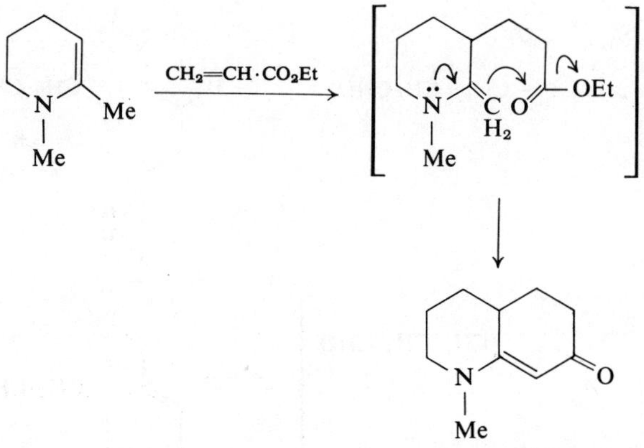

$CH_2{=}CH \cdot CO_2Et$

With nitro-olefines. The reaction between ketone enamines and nitro-olefines can be quite complex. The product formed depends very largely upon the experimental conditions as well as upon steric and electronic factors. The products of scheme 2.18 are typical examples.

Scheme 2.18

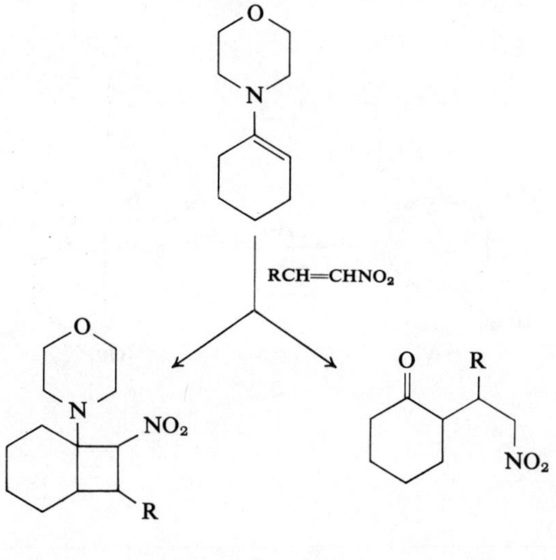

$RCH{=}CHNO_2$

2.1.3b. Enamines of aldehydes. Simple alkylation products can be obtained, especially from those enamines that possess a hydrogen atom attached to the β-carbon atom. Usually, however, the initial zwitterion collapses to a neutral species by internal neutralisation and cyclobutane derivatives are formed:

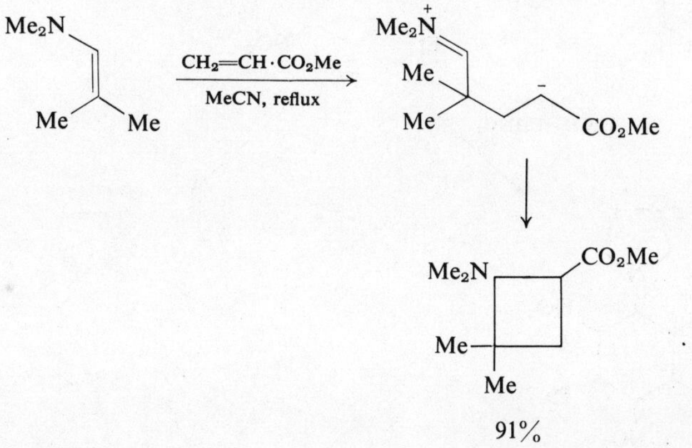

Similar products can be obtained from α,β-unsaturated nitriles and from nitro-olefines. If the initial enamine possesses a β-hydrogen atom, the cyclobutane formed is thermally unstable and reverts to starting materials or simple alkylation products or to both of these:

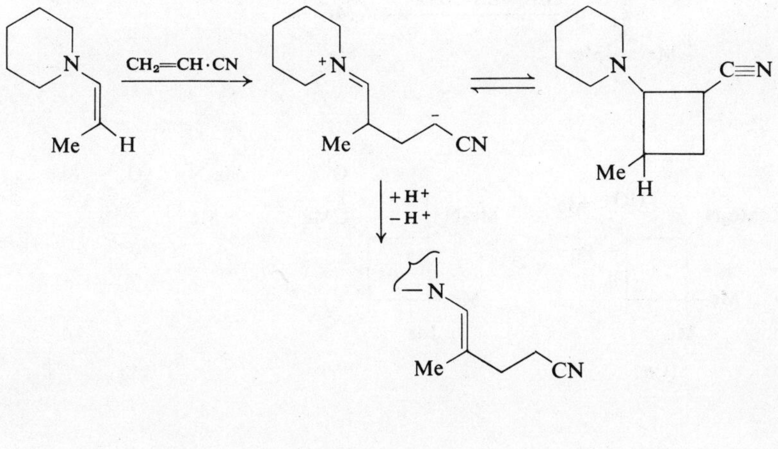

Apart from the cyclobutane derivatives, nitro-olefines can yield a different type of product in which two molecules of nitro-olefine and one molecule of enamine are involved:

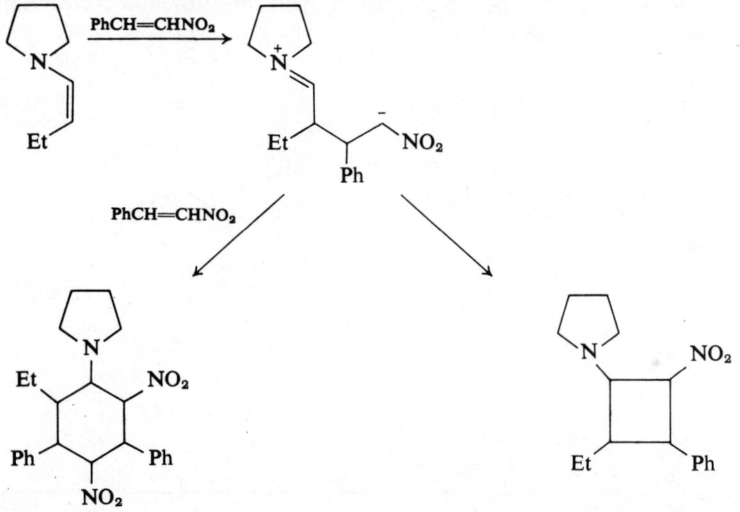

$$\xrightarrow{PhCH{=}CHNO_2}$$

$PhCH{=}CHNO_2$

α,β-unsaturated ketones give rise to dihydropyrans, but the reactions of the product indicate that it may be in equilibrium with the cyclobutane isomer. For example, when the adduct (LV) is treated with phenyl lithium, the product is (LVI) (scheme 2.19).

Scheme 2.19

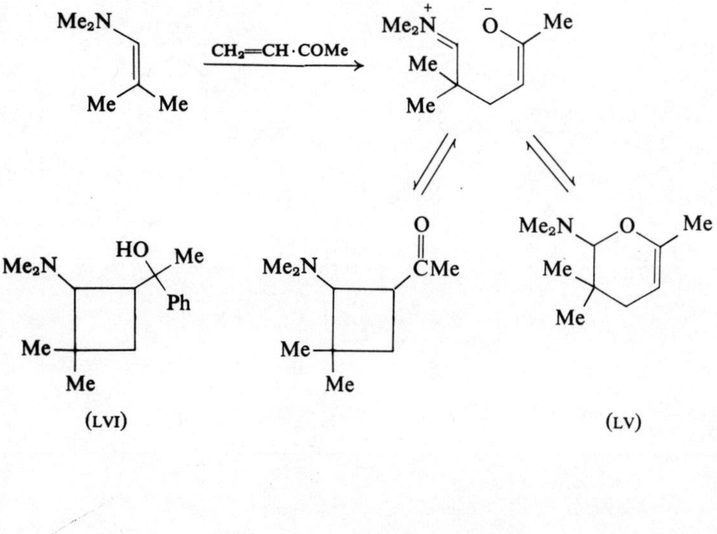

$$\xrightarrow{CH_2{=}CH\cdot COMe}$$

(LVI) (LV)

α,β-unsaturated aldehydes react with aldehyde enamines to yield substituted glutaraldehydes. The reaction may be a straightforward alkylation:

Me_2N Me Me $\xrightarrow{CH_2{=}CH\cdot CHO}$ $Me_2\overset{+}{N}$ Me Me CHO ⟶ OHC CHO Me Me

or may involve an initial cycloaddition to form a dihydropyran which subsequently undergoes a rearrangement:

Me_2N Me Me ⟶ Me_2N O Me Me $\xrightarrow{\Delta}$ Me_2N CHO Me Me ⟶ OHC CHO Me Me

2.1.4. Miscellaneous electrophilic reagents. A large number of electrophilic reagents other than those already discussed react with enamines. The following section summarises some of these.

Arylation. Aryl halides that are activated towards nucleophilic reagents (e.g. 2,4-dinitrochlorobenzene) will react with enamines, but since they are less reactive than, for example, alkyl or allyl halides, it is important to use reactive enamines, such as those derived from pyrrolidine. Morpholine enamines do not react.

N + Cl NO_2 NO_2 $\xrightarrow{Et_3N}$ O NO_2 NO_2 $\xrightarrow{\text{reduction}}$ H_2N N H

With the less reactive halides such as *p*-nitrochlorobenzene, *N*-arylation occurs:

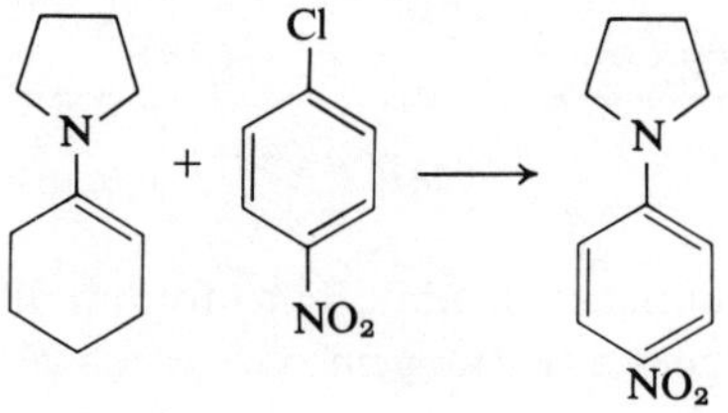

Enamines have been reported to react with benzyne (generated from *o*-bromo-fluorobenzene and magnesium):

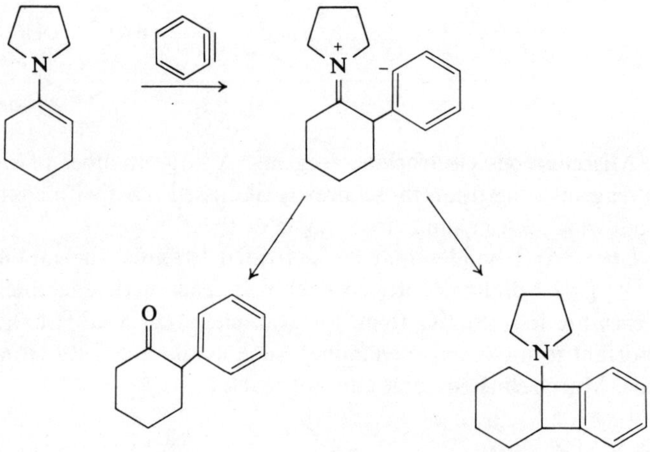

Reaction with aldehydes. Good yields of 2-alkylidene ketones are easily achieved by reaction of an enamine with an aldehyde, especially an aromatic aldehyde (scheme 2.20).

Scheme 2.20

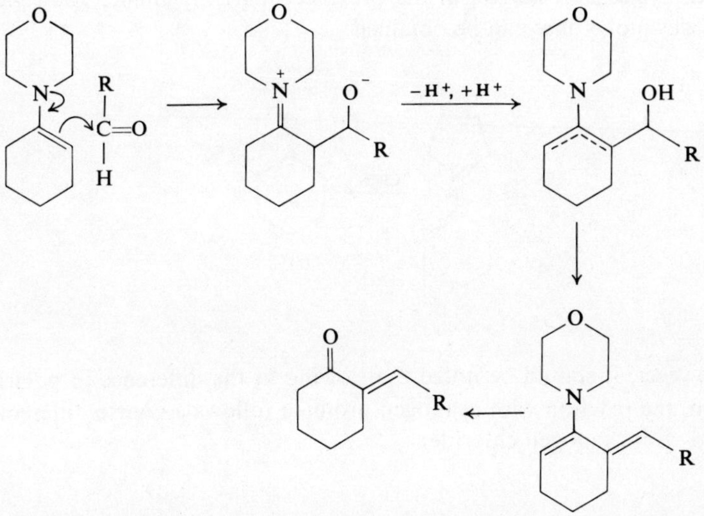

R, C=O, H, O, N, $\overset{+}{\text{N}}$, O^-, R $\xrightarrow{-H^+,\ +H^+}$ OH, R

Reaction with halogens. Chlorination and bromination reactions of enamines have been described, yielding the β-haloiminium salt, which is readily hydrolysed. Stable β-haloenamines have been isolated when *N*-halosuccinimides were used as halogenating agents:

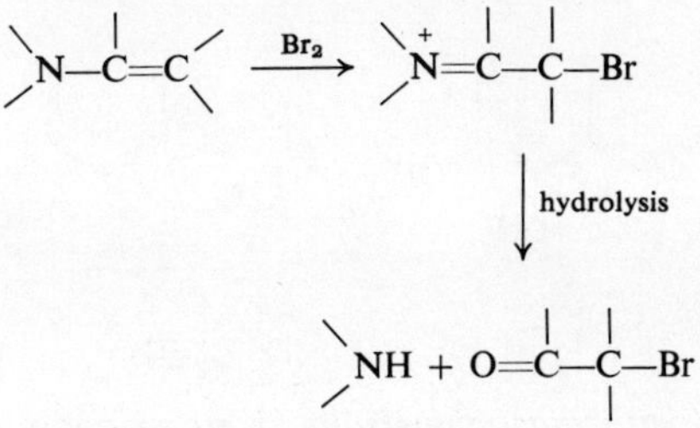

$$\text{N-C=C} \xrightarrow{\ \text{Br}_2\ } \overset{+}{\text{N}}\text{=C-C-Br} \xrightarrow{\text{hydrolysis}} \text{NH} + \text{O=C-C-Br}$$

and

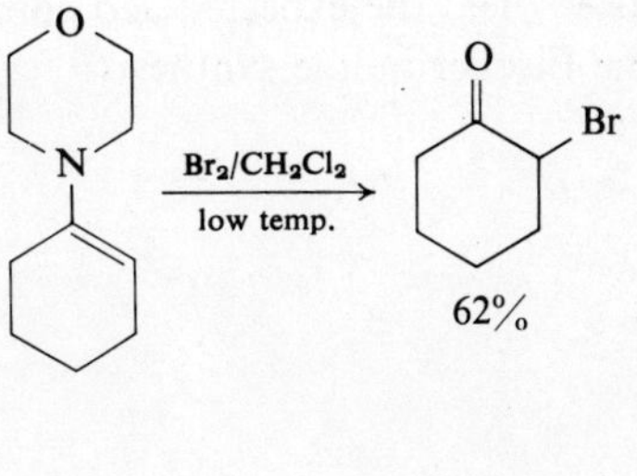

$$\xrightarrow[\text{low temp.}]{\text{Br}_2/\text{CH}_2\text{Cl}_2}$$

62%

Reaction with cyanogen halides. When pyrrolidine enamines react with cyanogen *chloride*, in the presence of triethylamine, good yields of α-cyanoketones can be obtained:

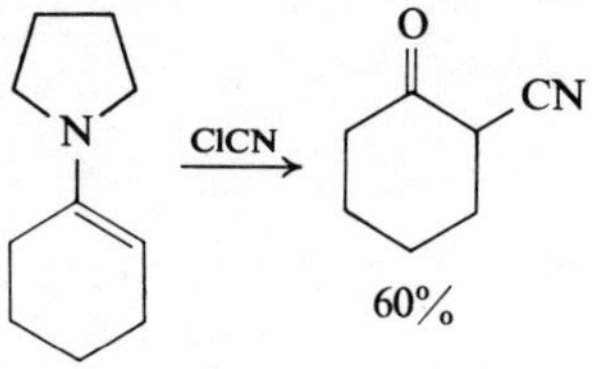

ClCN

CN

60%

However, it should be noted that, owing to the difference in polarisation, the reaction with cyanogen bromide follows a course different to that with cyanogen chloride:

BrCN

Br

ClCN

OH

CN

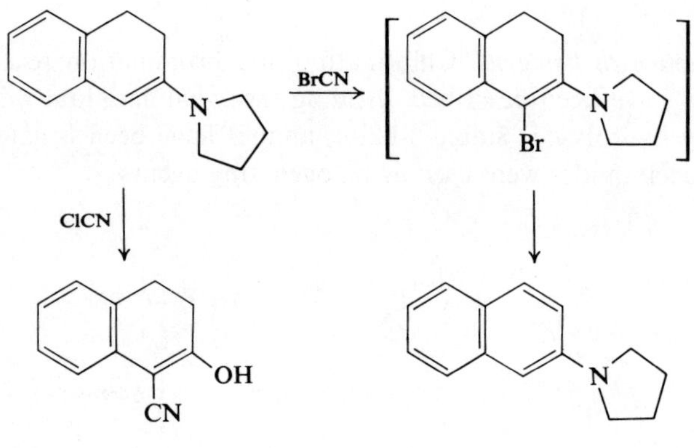

Reaction with diazonium salts. Ketone enamines react readily with diazonium salts to yield the expected azo compound; this reaction has been used in the Fischer indole synthesis:

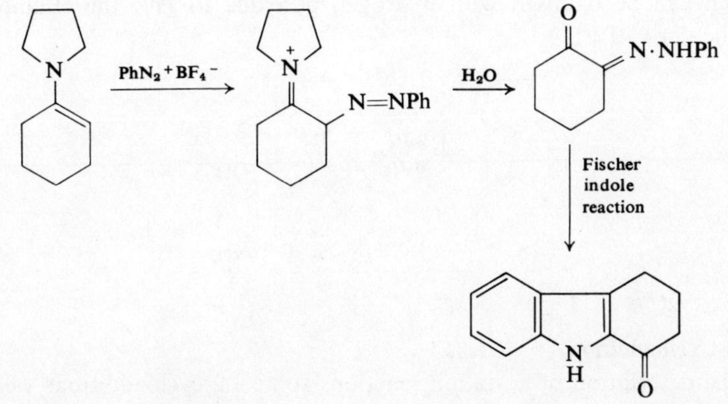

N $\xrightarrow{PhN_2^+BF_4^-}$ $\overset{+}{N}$ N=NPh $\xrightarrow{H_2O}$ O N·NHPh

Fischer indole reaction

N H O

The course of the reaction with aldehyde enamines depends upon whether the enamine contains a β-hydrogen atom:

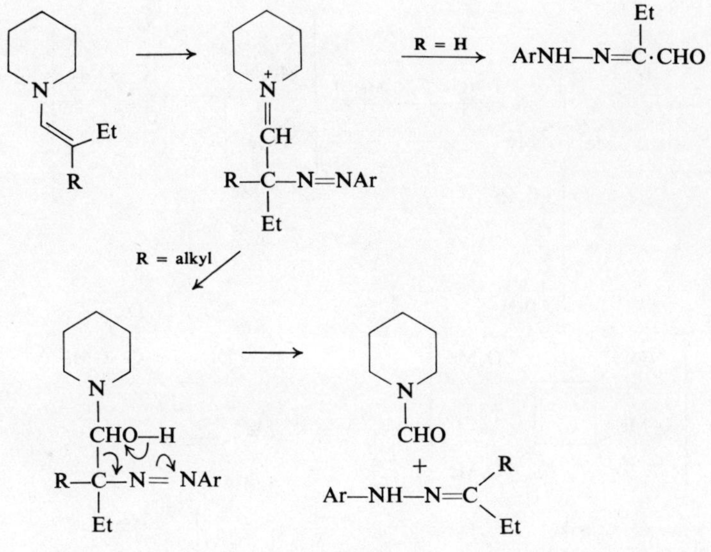

N Et R $\longrightarrow$ $\overset{+}{N}$ CH R—C—N=NAr Et $\xrightarrow{R\,=\,H}$ ArNH—N=C(Et)·CHO

R = alkyl

N CHO—H R—C—N=NAr Et $\longrightarrow$ N CHO + Ar—NH—N=C(R)(Et)

Hydroboration. Diborane reacts with enamines to form the amino borane:

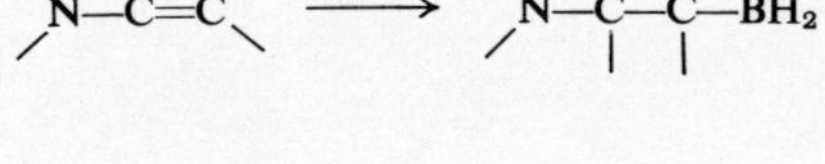

$$\rangle N{-}C{=}C\langle \xrightarrow{B_2H_6} \rangle N{-}C{-}C{-}BH_2$$

which can be oxidised with hydrogen peroxide to give the ethanolamine derivative:

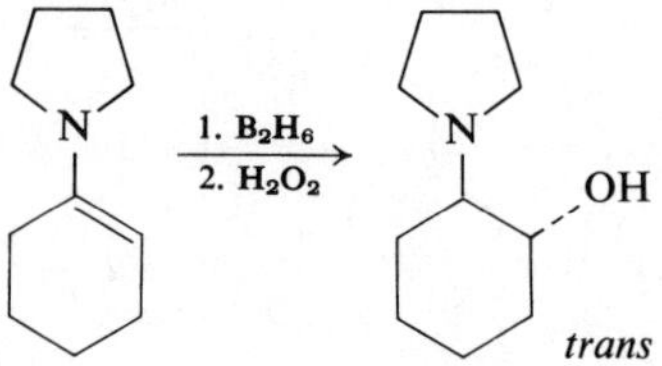

1. B_2H_6
2. H_2O_2
OH
trans

2.2. Cycloaddition

In the description of acylation reactions some 1,2-cycloadditions were seen to occur, whereas the reactions of enamines with electrophilic olefines involved 1,2-, 1,3- and 1,4-cycloadditions. Such reactions can

Scheme 2.21

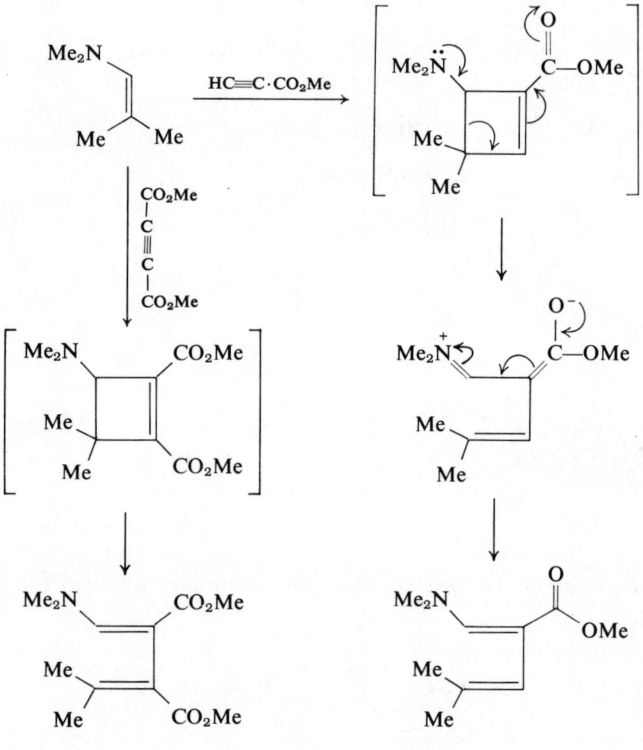

Me_2N Me Me $HC{\equiv}C{\cdot}CO_2Me$ $Me_2\ddot{N}$ O C—OMe Me Me

CO_2Me C ≡ C CO_2Me

Me_2N CO_2Me Me Me CO_2Me $Me_2\overset{+}{N}$ O^- C—OMe Me Me

Me_2N CO_2Me Me Me CO_2Me Me_2N O OMe Me Me

provide very attractive synthetic routes to a variety of structures as illustrated above. Some further cycloaddition reactions of synthetic interest are summarised in this section.

2.2.1. Reaction with acetylenes. Activated acetylenes react with enamines to yield cyclobutene derivatives that may then undergo a rearrangement reaction. The overall result with aldehyde enamines (scheme 2.21) is the insertion of a two carbon fragment into the enamine, whereas the reaction offers a method of ring-expansion by two carbon atoms when enamines of cyclic ketones are used (scheme 2.22).

Scheme 2.22

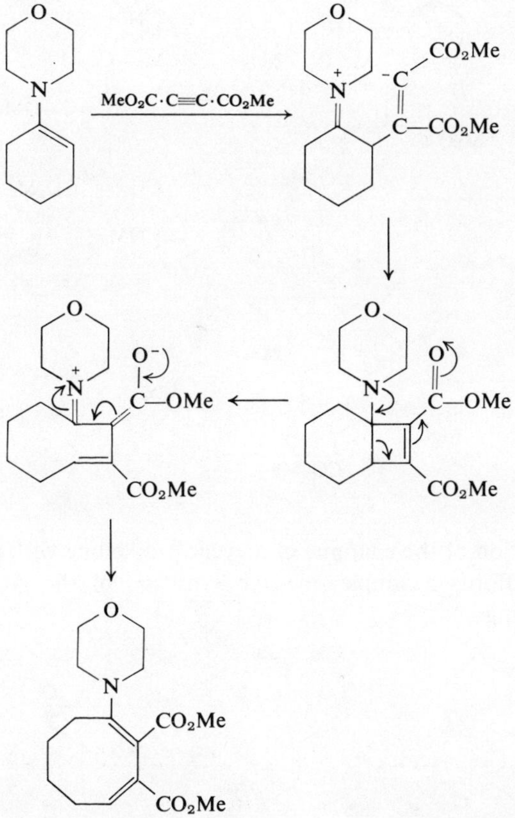

O N $MeO_2C \cdot C \equiv C \cdot CO_2Me$ O $\overset{+}{N}$ ^-C CO_2Me $C—CO_2Me$

O N O C—OMe CO_2Me

O $\overset{+}{N}$ O^- C—OMe CO_2Me

O N CO_2Me CO_2Me

If the enamine of an acyclic diketone is used, an interesting synthesis of the benzene ring is possible:

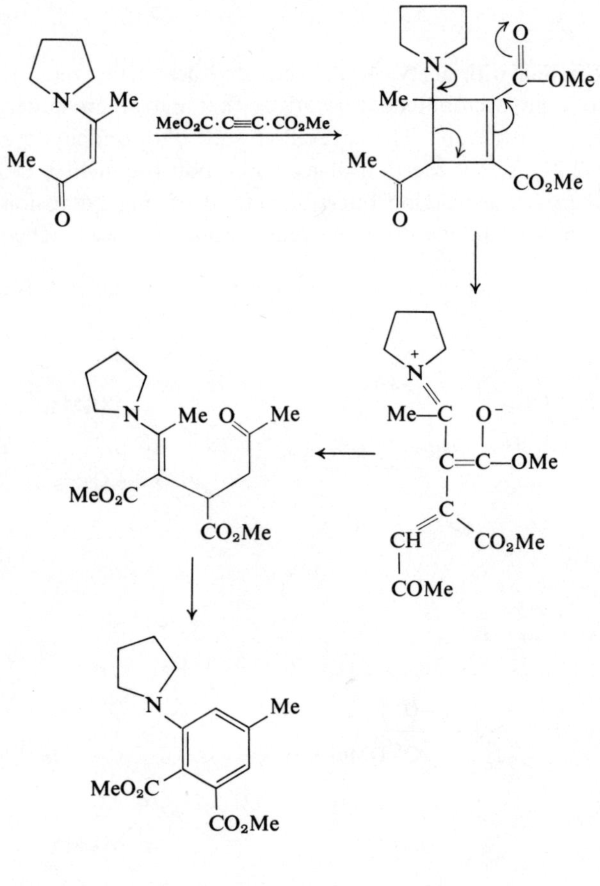

The reaction of the enamine of a cyclic β-diketone with an activated acetylene affords a simple one-step synthesis of the Ar-tetrahydro-quinoline ring:

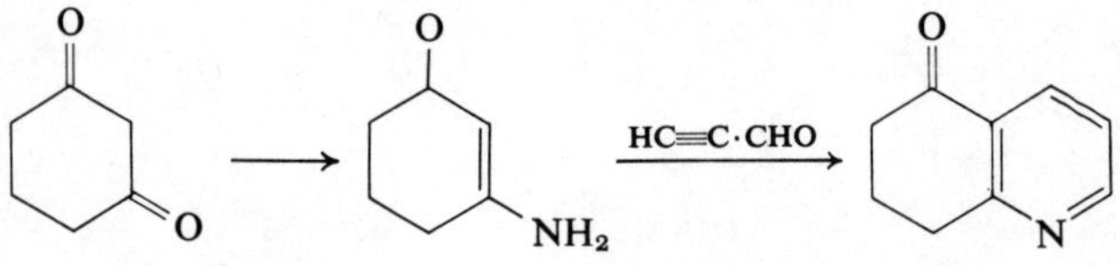

2.2.2. The Diels–Alder reaction. In the normal Diels–Alder reaction, an electron-rich diene undergoes a 1,4-cycloaddition reaction with an electron-poor dienophile:

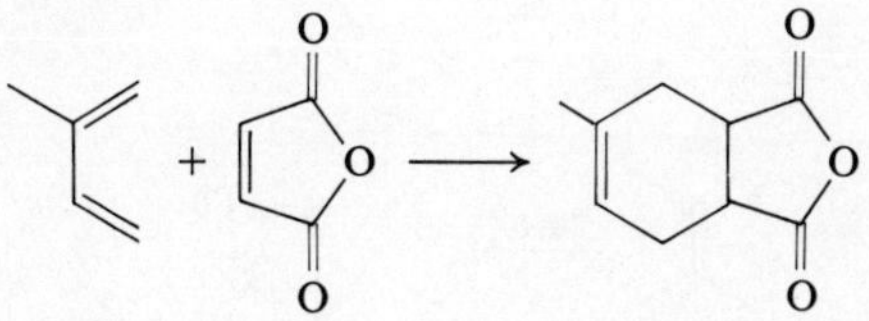

O O
\+ O ⟶ O
O O

Enamines can behave as dienophiles, but since they contain an electron-rich double bond, the diene must be electron-deficient. Methyl *trans*-2,4-pentadienoate has been shown to react in this way with several enamines:

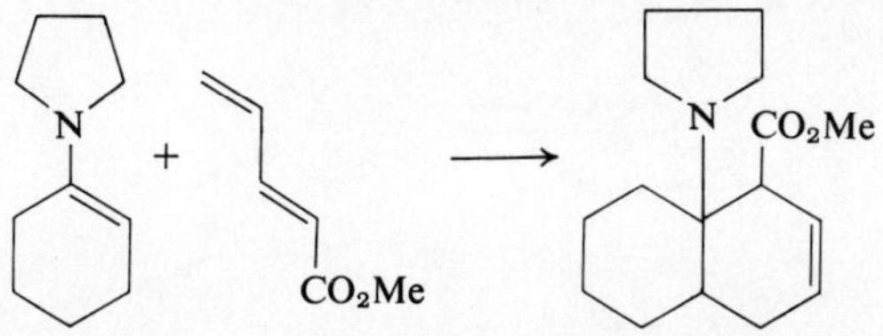

N + ⟶ N CO_2Me
CO_2Me

and

CO_2Me
N ⟶ MeN
N
Me

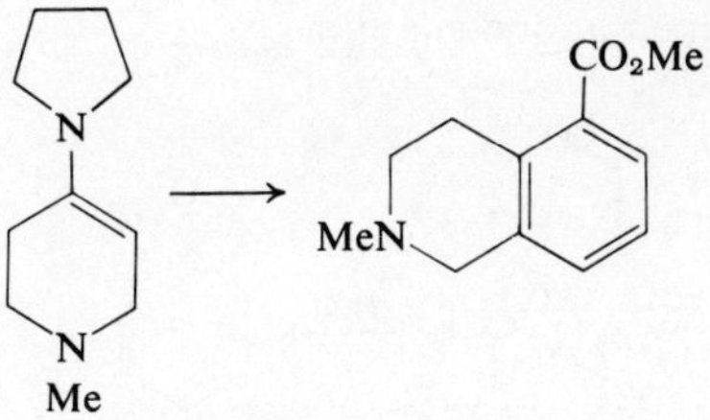

However, dienamines may be employed as electron-rich dienes and as such will react with the usual electrophilic olefines:

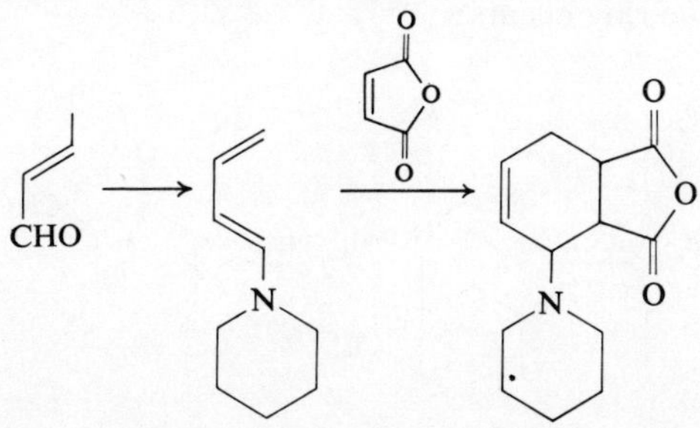

O
O O
O O
CHO ⟶ ⟶
N N O

2.2.3. 1,3-Dipolar additions. Enamines will behave as the dipolarophile in 1,3-cycloaddition reactions, involving such partners as nitrones, nitrile oxides and azides:

N PhN_3 N Ph N N N

N Ph—C≡N^{+}—O^{-} N O N Ph

2.2.4. Reaction with carbenes. This reaction can be employed to expand the ring with cyclopentanone enamines:

O N $:CCl_2$ O N CCl_2 H_2O O Cl

but with cyclohexanone enamines, an alternative mode of opening of the cyclopropane ring occurs:

O N $:CCl_2$ O N CCl_2 H_2O O CHCl

2.2.5. Other cycloadditions. A considerable number of 1,4-cycloadditions involving bifunctional molecules have been reported. The following is a typical example:

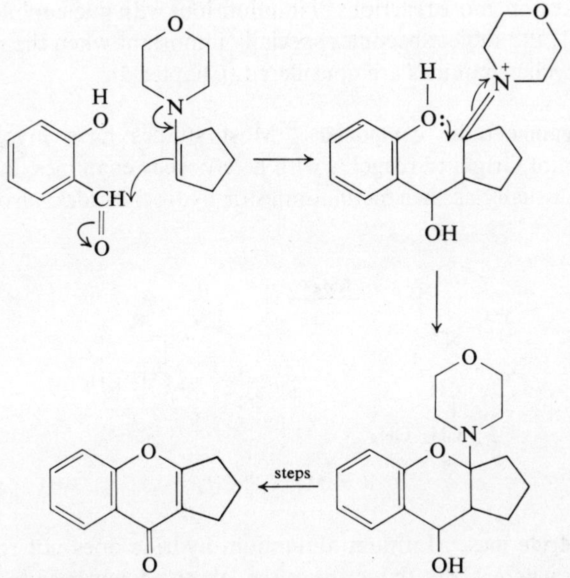

steps

2.3. Nucleophilic reagents

The structure of the iminium ion moiety can be regarded as a resonance hybrid of the canonical forms (LVII) and (LVIII). The reactivity of such structures has been compared with that of the carbonyl group; the chemistry is dominated by attack of nucleophiles at the electron-deficient carbon centre. Examples of such reactions have already been described

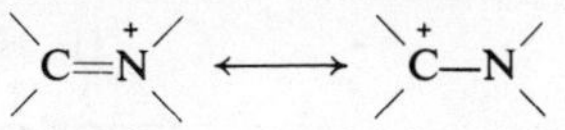

$$>C=\overset{+}{N}< \longleftrightarrow >\overset{+}{C}-N<$$

(LVII) (LVIII)

$$>C=O \longleftrightarrow >\overset{+}{C}-O^-$$

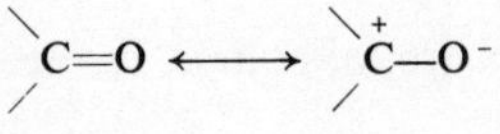

as incidentals to the hydrolysis of enamines and to the various cycloaddition reactions described in this chapter.

Iminium ions are generated when the enamine reacts with a variety of electrophiles, including the proton. Methods of detection of such ions were mentioned in Chapter 1 when the ultraviolet and infrared spectra of enamine derivatives were described.

In this section, more reactions of iminium ions with nucleophiles will be mentioned; the subject becomes specially important when the reactions of heterocyclic enamines are considered (Chapter 3).

2.3.1. Organometallic compounds. Most studies have involved the interaction of Grignard reagents with heterocyclic enamines, but simple enamines, usually as their perchlorates or hydrochlorides, also react:

$$\overset{+}{N}(\text{piperidinium})=CH-CH_2\cdot CH_3 \xrightarrow{RMgX} N(\text{piperidino})-CH(R)-CH_2\cdot CH_3$$

R = Me or $PhCH_2$

2.3.2. Hydride ions. Lithium aluminium hydride does not react with enamines, whereas the reduction often observed when sodium borohydride is used is due to the protonation of the enamine by the solvent (aqueous methanol) and reduction of the iminium ion thus generated:

$$>N-C=C< \;\longrightarrow\; >\overset{+}{N}=C-CH< \;\xrightarrow{H^-}\; >N-CH-CH<$$

Pre-formed iminium ions are, of course, reduced by lithium aluminium hydride.

2.3.3. Reaction with diazoalkanes. The addition of diazomethane to the perchlorates of enamines leads to aziridinium salts:

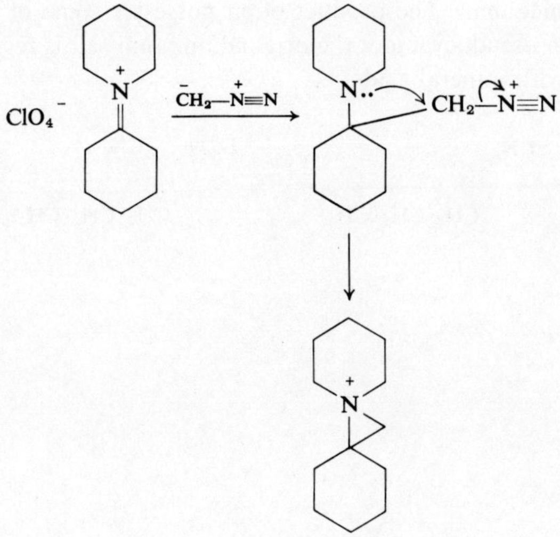

When these salts are treated with other nucleophilic reagents, e.g. Cl^-, ring-opening occurs:

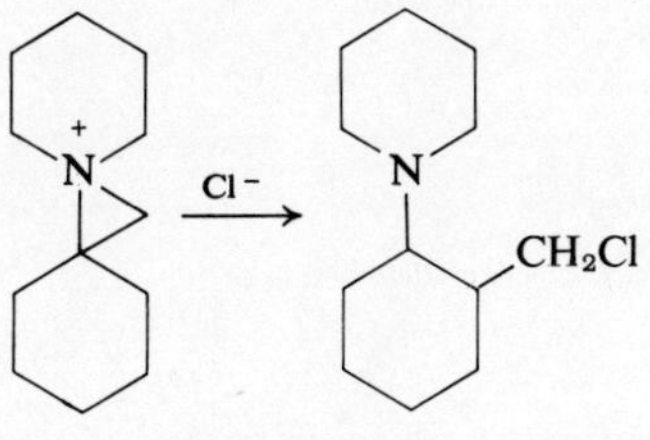

When diazoethane is used in place of diazomethane, the expected product (LIX) is accompanied by the ring-expanded enamine (LX):

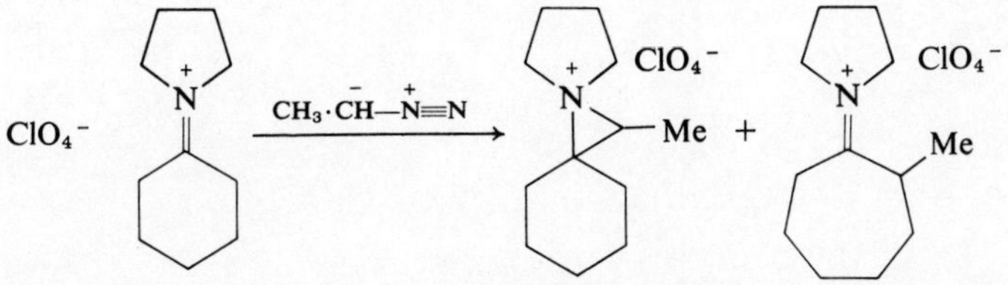

(LIX) (LX)

2.3.4. Cyanide ion. The product often possesses some of the characteristics of a pseudocyanide; the original iminium salt is regenerated by treatment with mineral acids:

$$Et_2\overset{+}{N}{=}CH{-}CH_2{\cdot}CH_2{\cdot}CH_3 \underset{H^+}{\overset{KCN}{\rightleftharpoons}} Et_2N{-}CH(CN){-}CH_2{\cdot}CH_2{\cdot}CH_3$$

3 Heterocyclic enamines

3.1. Definition

The term heterocyclic enamine is used to describe those compounds that possess the >C=C—N< group as part of the ring system. Simple examples are Δ^2-pyrrolines (I), Δ^2-piperideines (II), indoles (III), 1,4-dihydroquinolines (IV) and 1,2-dihydroisoquinolines (V), together

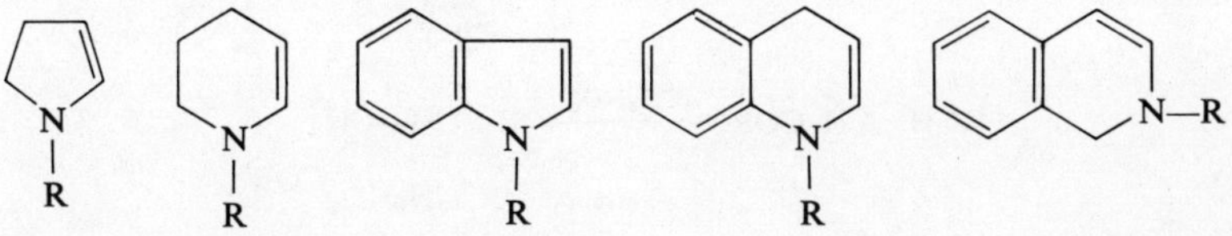

(I) (II) (III) (IV) (V)

with some fused bicyclic compounds such as (VI), (VII) and (VIII). This chapter will be confined almost exclusively to these types of system, although mention will also be made of the closely related exocyclic derivatives such as (IX).

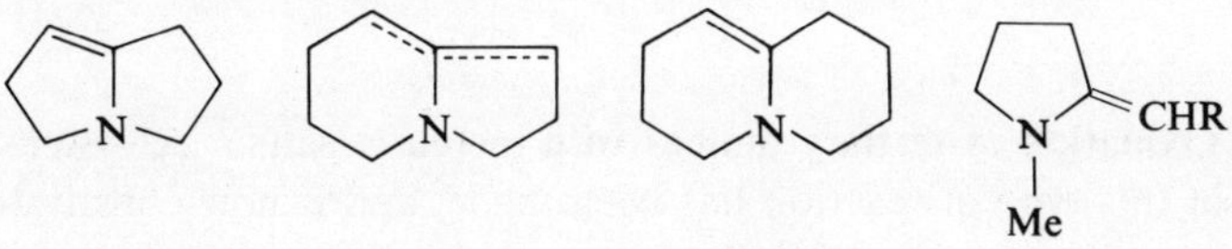

(VI) (VII) (VIII) (IX)

3.2. Preparation

3.2.1. Condensation of a carbonyl compound with a secondary amine. This standard method of enamine formation has only limited, specialised application in the preparation of heterocyclic enamines (scheme 3.1).

Scheme 3.1

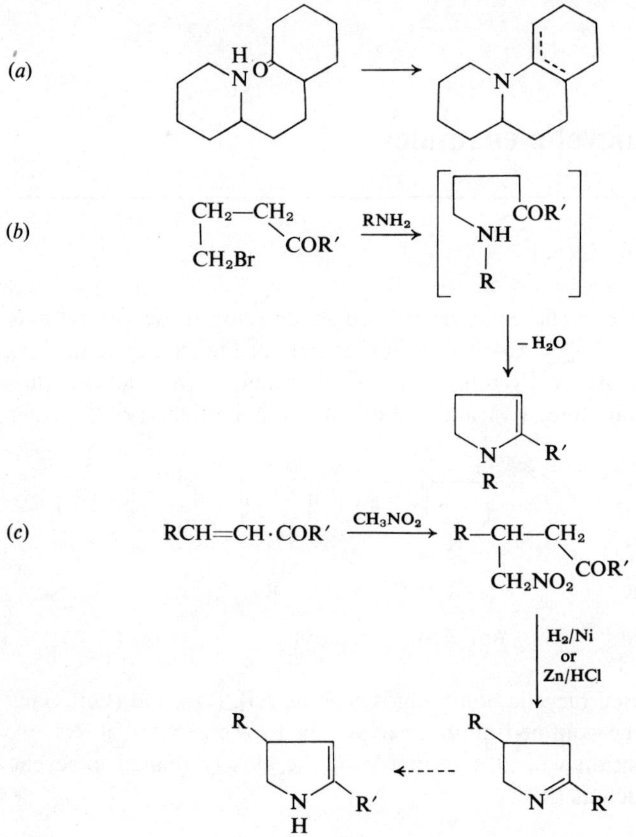

(*a*) H N O → N

(*b*) $CH_2—CH_2$ | CH_2Br COR' $\xrightarrow{RNH_2}$ [COR' NH | R] $\xrightarrow{-H_2O}$ N R R'

(*c*) $RCH{=}CH·COR'$ $\xrightarrow{CH_3NO_2}$ $R—CH—CH_2$ | CH_2NO_2 COR' $\xrightarrow{H_2/Ni \text{ or } Zn/HCl}$ R N R' ←----- R N H R'

3.2.2. Oxidation of tertiary amines with mercuric salts. A systematic study of this type of reaction has been made, and it now constitutes a valuable general method for the preparation of certain types of enamines. In the first example to be reported, a mixture of quinolizidine and mercuric acetate, in dilute aqueous acetic acid solution was merely heated on a steam bath until the precipitation of mercurous acetate was complete:

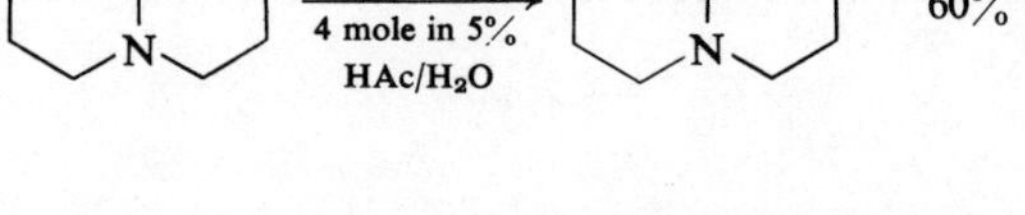

N $\xrightarrow[\text{4 mole in 5\% HAc/}H_2O]{Hg(OCOMe)_2}$ N 60%

When the two rings are unsymmetrically substituted, a mixture of products is usually obtained in which the isomer with the more highly substituted double bond predominates:

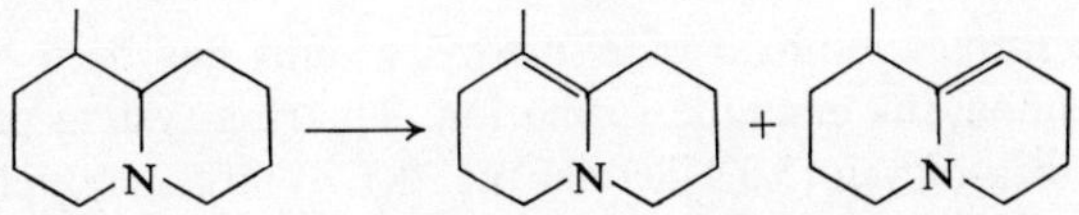

ratio of 2 : 1

It is a general rule, implicit in the above examples, that where a choice exists, a tertiary hydrogen atom that is attached to the carbon atom adjacent to nitrogen is removed in preference to a secondary α-hydrogen atom.

By a close study of a variety of bicyclic amines, and by the use of deuterium-labelled derivatives, it has been established that the oxidation by mercuric acetate involves the initial formation of a mercurated complex, through the unshared electron pair on nitrogen. This is followed

Scheme 3.2

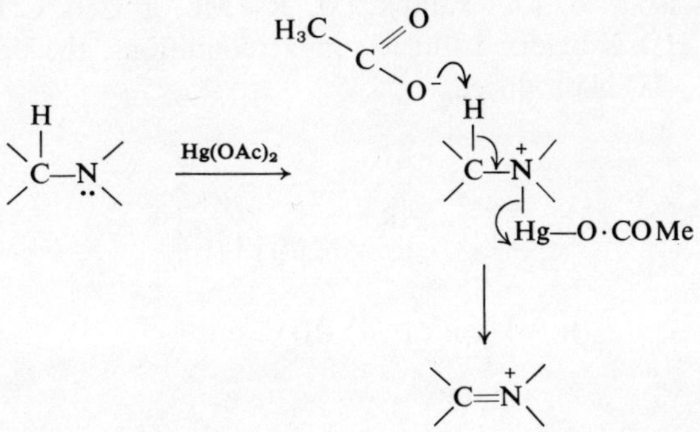

H_3C C O O^- H

H C—N $Hg(OAc)_2$ C—$\overset{+}{N}$ Hg—O·COMe

C=$\overset{+}{N}$

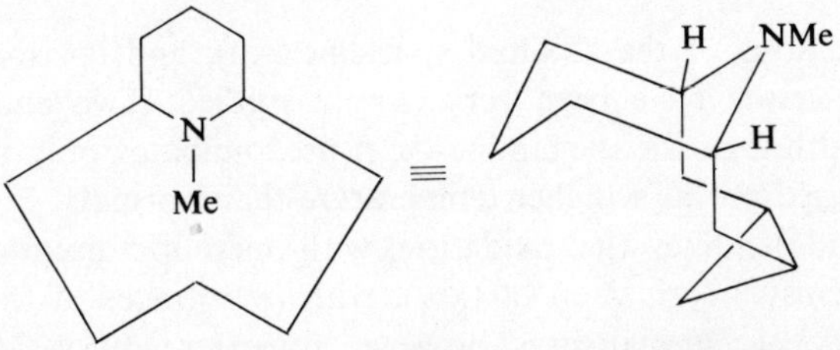

N Me ≡ H NMe H

(x)

by a concerted removal of a proton from the α-carbon atom and cleavage of the nitrogen–mercury bond (scheme 3.2). A *trans*-coplanar relationship should exist between the hydrogen atom being removed and the nitrogen–mercury complex.

In some examples removal of *N*-methyl groups has been observed rather than endocyclic enamine formation. The ring-system present in 11-methyl-11-aza-bicyclo[5.3,1]hendacane (x) is large enough to accommodate a double bond at the bridgehead should it be formed, but since the two, equivalent tertiary α-hydrogen atoms are not anti-parallel to the $>$N—Hg bond in the complex, neither is removed. Instead, the freely-rotating methyl group can be arranged so that one of its hydrogen atoms becomes *trans* to the $>$N—Hg bond; oxidation can thus occur:

$$>N{-}CH_3 \longrightarrow >\overset{+}{N}{=}CH_2 \xrightarrow{H_2O} >NH$$

The iminium ion is then hydrolysed to the secondary amine in the reaction medium. In some cases where the iminium ion formed in the oxidation cannot be converted into the endocyclic enamine, it is the iminium ion that is isolated, for example (XI, R=Me or $CH_2 \cdot CH_5$). If (XI, R=$CH_2 \cdot CH_3$) is oxidised under similar conditions, the de-alkylated base (XI, R=H) is produced.

$$NR \longrightarrow \overset{+}{N}R$$

R = Me or $CH_2 \cdot C_6H_5$

(XI)

The reactions of the alkaloid sparteine (XII), and its isomers, with mercuric acetate have been very closely studied. Two enamines are possible, although the dienamine (XIII) predominates only if the reaction is carried out at a higher temperature than normal.

With indolizidine, the oxidation with mercuric acetate leads to almost exclusive formation of (XIV), with only traces of the isomeric product (XV). In special cases, however, for example (XVI), the second isomer becomes the preferred one.

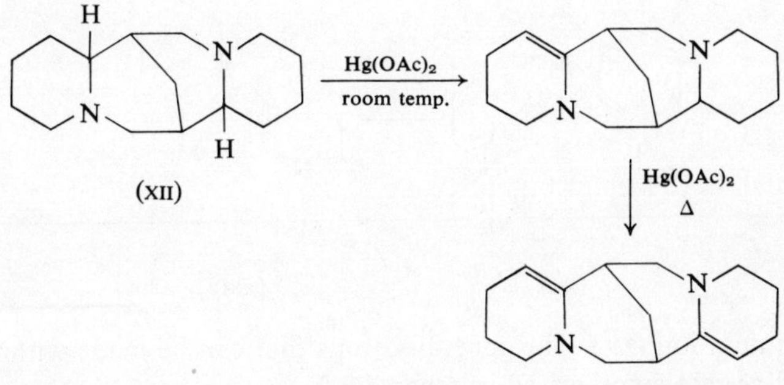

$Hg(OAc)_2$, room temp.

(XII)

$Hg(OAc)_2$, Δ

(XIII)

$Hg(OAc)_2$

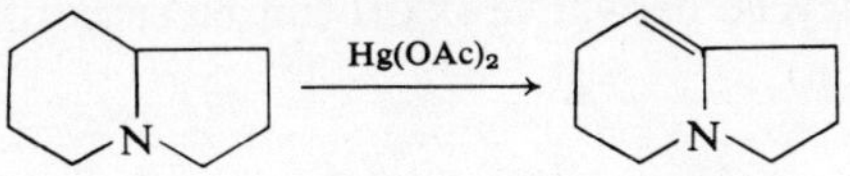

(XIV)

In bicyclic amines of the type (XVII), where the nitrogen is adjacent to the bridgehead, oxidation proceeds normally to give an enamine (XVIII), provided that the other bridgehead position is substituted (R≠H). Without such a substituent, the product is the hydroxylated enamine (XIX).

(XV) (XVI)

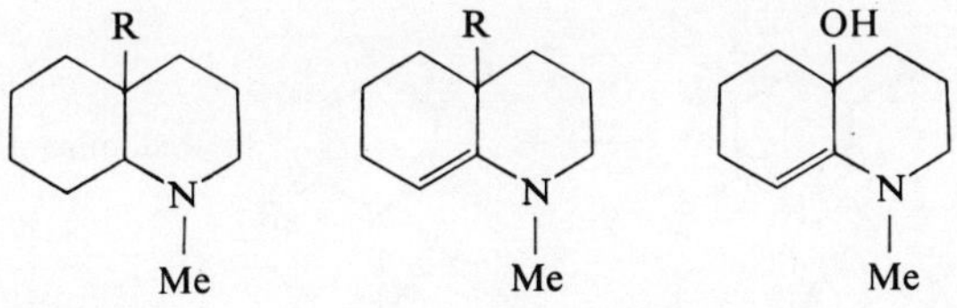

(XVII) (XVIII) (XIX)

An extensive study has been made of the oxidation of piperidines with mercuric acetate. Piperidine itself is resistant to oxidation, but the *N*-methyl compound is easily oxidised to the dimer (XXI), presumably via

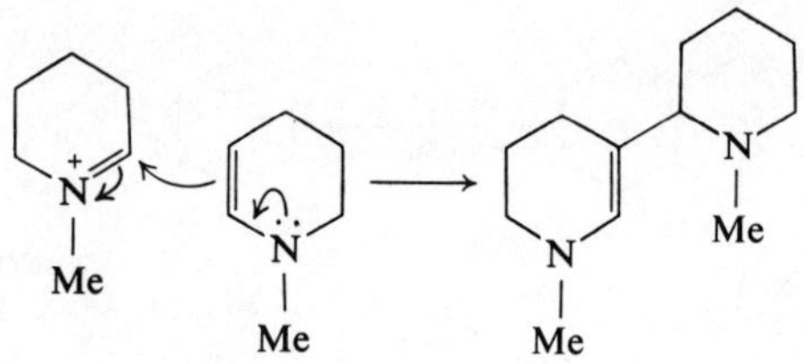

(XX) (XXI)

the iminium ion (XX). The generalisations that can be made with other piperidines are summarised in scheme 3.3.

The partial dehydrogenation of *N*-alkyl pyrrolidines with mercuric acetate is not a preparative method since the reaction is complex. However, the bicyclic derivative (XXII) can be successfully oxidised to the enamine (XXIII).

Scheme 3.3

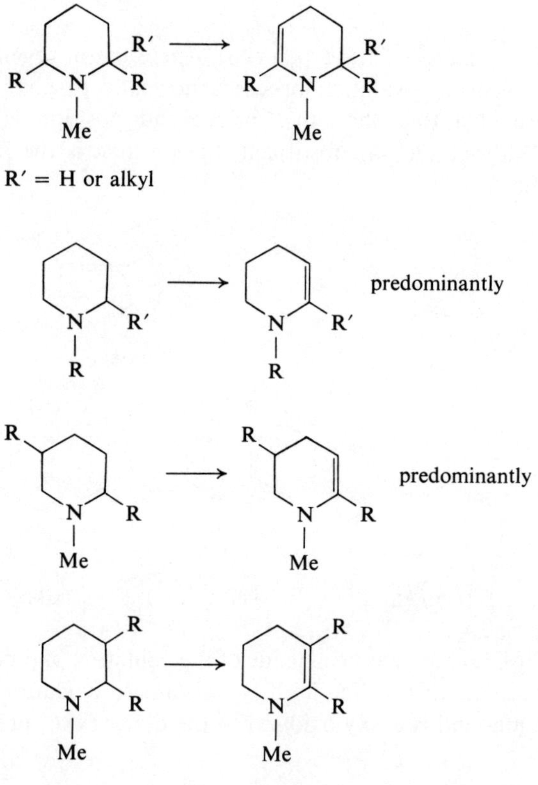

R′ = H or alkyl

predominantly

predominantly

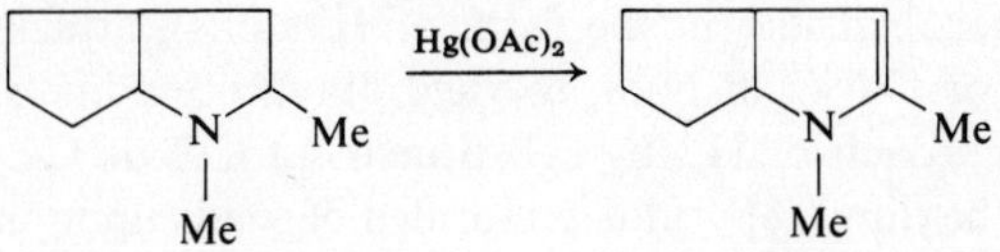
Hg(OAc)2
N
Me
Me
N
Me
Me

(XXII) (XXIII)

3.2.3. Partial reduction of aromatic nitrogen heterocycles

3.2.3a. An intensive study has been made of the partial reduction of pyridine derivatives, with a wide variety of reagents (scheme 3.4). The reagents of choice, however, are sodium borohydride and lithium aluminium hydride (LAH).

Scheme 3.4

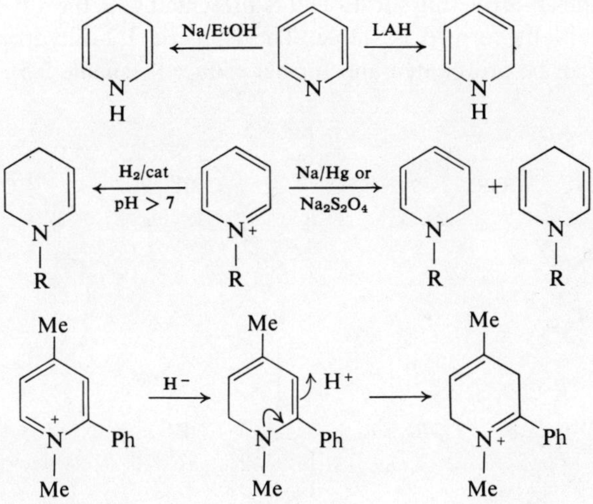
Na/EtOH
LAH
H2/cat
pH > 7
Na/Hg or
Na2S2O4
H−
H+
Ph
Me

(XXIV) (XXV) (XXVI)

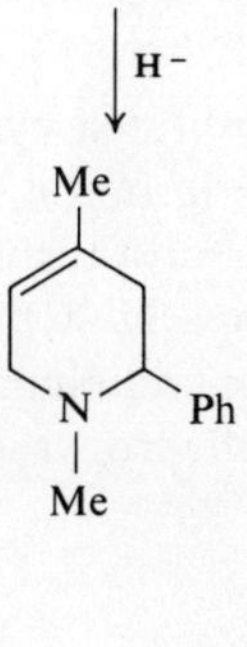
H−
Me
Ph
Me

(XXVII)

Sodium borohydride (in the form of H^- ions), attacks pyridinium salts at C-2 or C-6 or at both, to yield dihydropyridine(s). Steric hindrance to approach of H^- by substituents at C-2 or C-6 is common. The fate of the dihydropyridine(s) formed depends upon the nature and the positions of substituents attached to the ring. The pyridinium ion (XXIV) undergoes attack by H^- at C-6, and the resulting dihydropyridine (XXV) is protonated by the solvent at C-3 so that the resulting iminium ion (XXVI) can then be further reduced by an excess of sodium borohydride to the 1,2,3,6-tetrahydropyridine (XXVII). If only a molar equivalent of reducing agent is used, or if the reduction is carried out in the presence of alkali, the 1,6-dihydropyridine can be isolated. An alternative procedure is to use an aprotic solvent such as dimethylformamide or 1,2-dimethoxyethane instead of the more usual aqueous ethanol, when the dihydropyridines can be isolated; this constitutes a good method for the preparation of such dienamines.

If an electron-attracting substituent is present at C-3, the 1,6-dihydropyridine initially formed is stable. The isomeric 1,2-dihydropyridine, however, can be protonated and further reduced (scheme 3.5).

Scheme 3.5

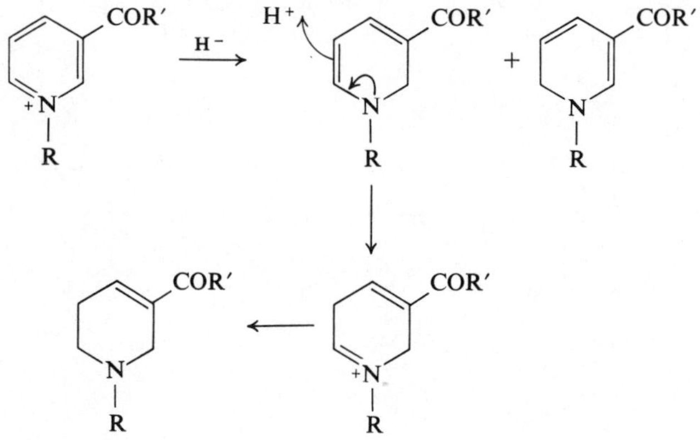

COR′ $\xrightarrow{H^-}$ H⁺ COR′ + COR′

+N–R N–R N–R

COR′ ⟵ COR′

N–R +N–R

3,5-Disubstituted pyridinium ions (XXVIII) are reduced with sodium borohydride in protic solvents only as far as the dihydro-stage (XXIX), because the substituent at C-3 sterically hinders the protonation step that is an essential prerequisite for further reduction. However, 2-substituted pyridinium salts (XXX, R=Me or CO_2Et) are reduced to 1,2,3,6-tetrahydropyridines (XXXI).

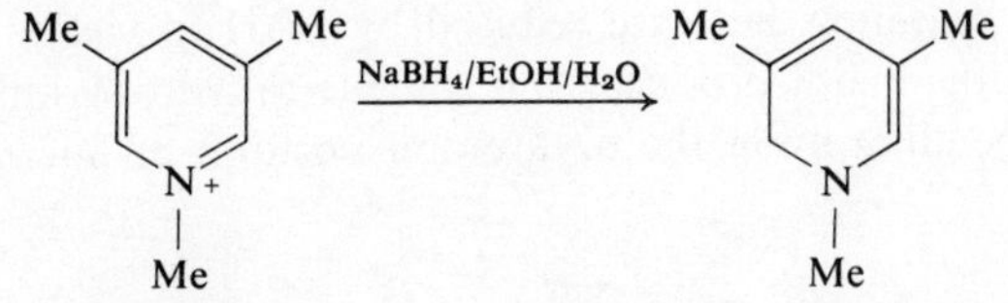

NaBH$_4$/EtOH/H$_2$O

(XXVIII) → (XXIX)

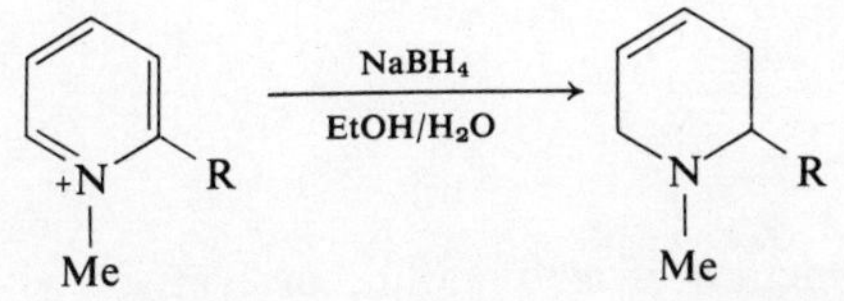

NaBH$_4$ EtOH/H$_2$O

(XXX) → (XXXI)

Pyridine itself is reduced by LAH to 1,2-dihydropyridine, whereas with methyl nicotinate, only the ester function is affected. Some other examples are given in scheme 3.6.

Scheme 3.6

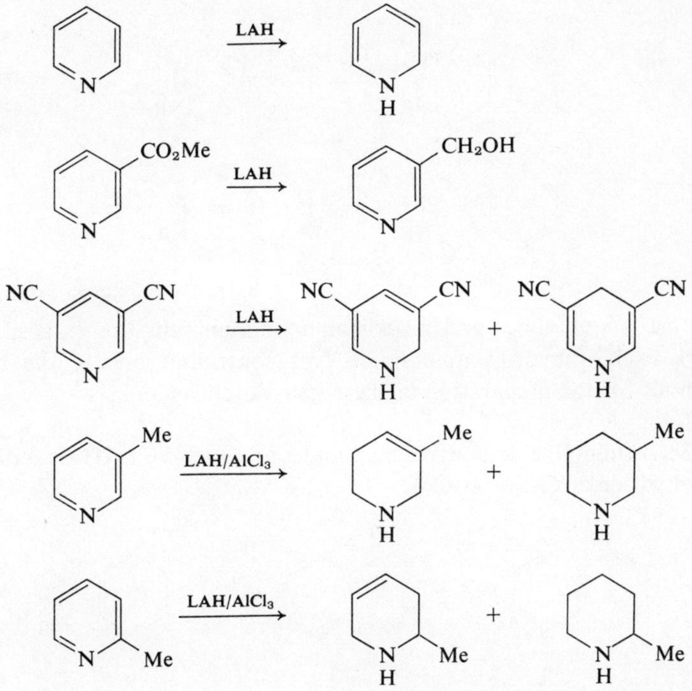

LAH

LAH

LAH

LAH/AlCl$_3$

LAH/AlCl$_3$

N-Alkylpyridinium salts are reduced by LAH to yield 1,2-dihydropyridines as the major products, but some tetrahydropyridines are also formed, depending upon the nature and position of nuclear substituents, e.g.

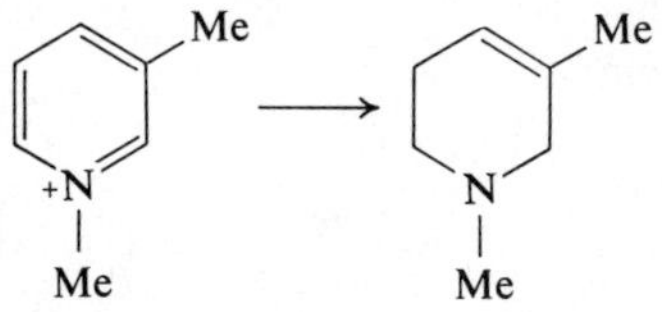

Me Me
+N N
Me Me

3.2.3b. The reduction of isoquinoline derivatives with sodium borohydride and with LAH has also been extensively studied. Isoquinolinium salts are usually reduced to the 1,2,3,4-tetrahydroisoquinoline derivatives with sodium borohydride, but in some cases, where an alkyl group is present at C-4, the 1,2-dihydroisoquinoline only is formed. This partial reduction is also observed if an aprotic solvent is used.

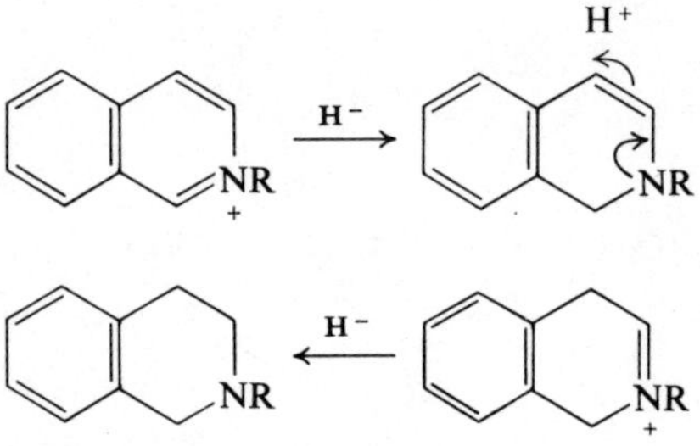

H^+
H^-
NR NR
+
H^-
NR NR
+

Some isoquinolines and most isoquinolinium salts are reduced by LAH to 1,2-dihydroisoquinolines. This constitutes one of the best methods for the preparation of these heterocyclic enamines.

3.2.3c. Quinoline derivatives may undergo attack by LAH or sodium borohydride at C-2 or at C-4:

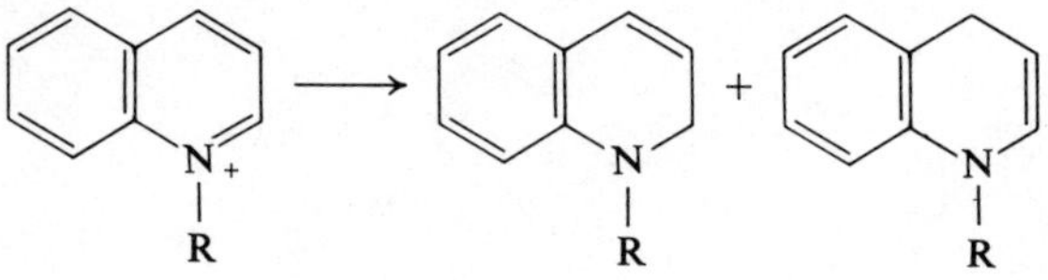

N+ N + N
R R R

With sodium borohydride, the intermediate dihydroquinoline is further reduced to the 1,2,3,4-tetrahydroquinoline derivative. However, when LAH is used, the 1,2-dihydroquinoline seems to predominate, although some isomeric 1,4-dihydroquinoline is also formed. The proportions of the isomers seem to depend upon the nature and position of substituents attached to the heterocyclic ring.

3.2.3d. The partial reduction of pyrroles with complex hydrides is not a useful method for the preparation of Δ^2-pyrrolines since complex products are often formed.

Indoles are not reduced by either LAH or sodium borohydride.

3.2.4. From lactams. Cyclic amides, which are often readily available, can be reduced by LAH to Δ^2-pyrrolines or Δ^2-piperideines (scheme 3.7) and this constitutes a useful method for the formation of such enamines. Alternatively, the lactam can react with Grignard reagents (scheme 3.7).

Scheme 3.7

(*a*) N-Me lactam (O) —LAH→ N-Me pyrroline

(*b*) N-Me lactam (O) —LAH→ [N-Me piperideine]

(*c*) NMe, O —LAH→ NMe

(*d*) N-Me lactam (O) —RMgX→ N-Me, R

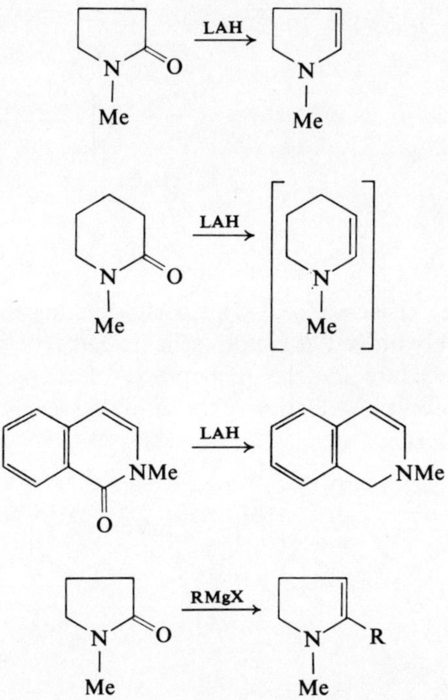

(e) [N-methyl-2-piperidone] $\xrightarrow{RMgX}$ [1-methyl-6-R-1,2,3,4-tetrahydropyridine]

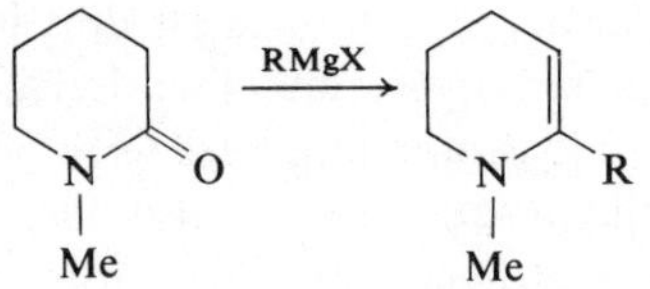

(f) [1,3,3-trimethyl-2-pyrrolidone] $\xrightarrow{R'CH_2MgX}$ [1,3,3-trimethyl-2-(R'-methylene)pyrrolidine]

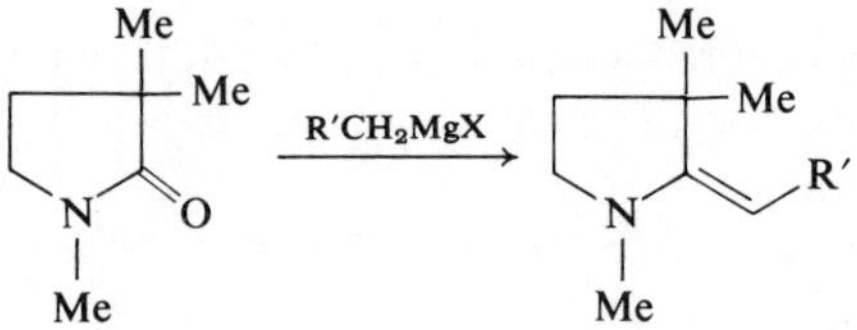

3.2.5. From nitriles. This method of approach to heterocyclic enamines, which seems to be rather limited in scope, is illustrated in scheme 3.8.

Scheme 3.8

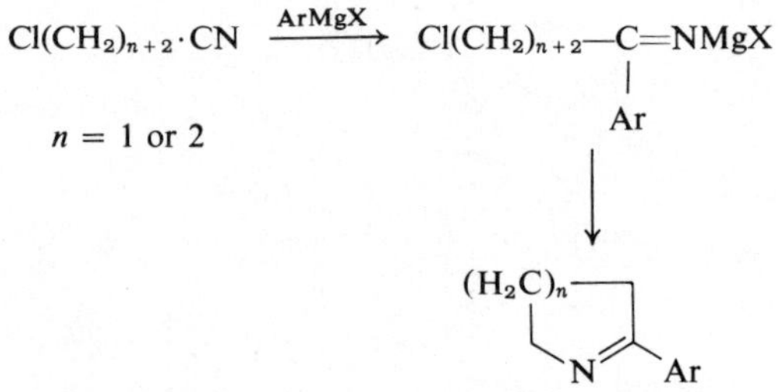

$$Cl(CH_2)_{n+2}\cdot CN \xrightarrow{ArMgX} Cl(CH_2)_{n+2}—C(Ar)=NMgX$$

$n = 1$ or 2

$$\downarrow$$

[cyclic imine: $(H_2C)_n$ ring, N=C–Ar]

3.2.6. By Claisen condensation. This useful-looking method seems to have received only limited attention (scheme 3.9).

The best procedure for the preparation of Δ^1-pyrroline and Δ^1-piperideine involves the oxidation of the saturated secondary amine with t-butyl hyperchlorite:

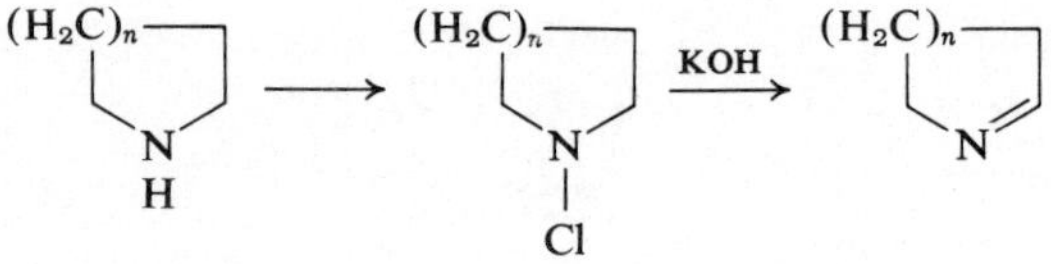

[$(H_2C)_n$ ring N–H] $\longrightarrow$ [$(H_2C)_n$ ring N–Cl] $\xrightarrow{KOH}$ [$(H_2C)_n$ ring N=C]

$n = 1$ or 2

Scheme 3.9

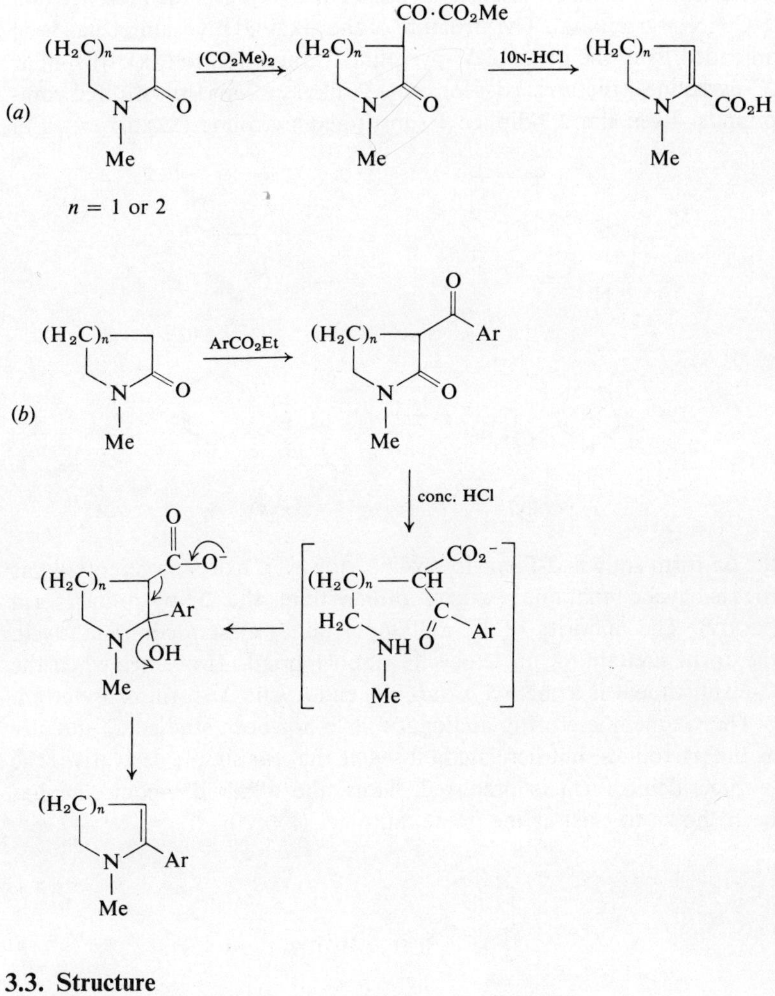

(*a*)

$(H_2C)_n$, N, O, Me $\xrightarrow{(CO_2Me)_2}$ $CO\cdot CO_2Me$, $(H_2C)_n$, N, O, Me $\xrightarrow{10N\text{-}HCl}$ $(H_2C)_n$, N, CO_2H, Me

$n = 1$ or 2

(*b*)

$(H_2C)_n$, N, O, Me $\xrightarrow{ArCO_2Et}$ O, Ar, $(H_2C)_n$, N, O, Me

conc. HCl

[CO_2^-, $(H_2C)_n$—CH, H_2C, C, O, Ar, NH, Me] ⟶ O, C, O^-, $(H_2C)_n$, Ar, N, OH, Me

⟶ $(H_2C)_n$, N, Ar, Me

3.3. Structure

3.3.1. Secondary amines. For the secondary amines, the equilibrium between enamine and imine tautomers, e.g.:

N, H ⇌ N

favours the latter (no band at about 3300 cm^{-1} in the infrared spectrum, but a band of medium intensity at 1620 cm^{-1} suggests the presence of a >C=N— grouping). The structure of the alkaloid myosamine has been amended from the original Δ^2-pyrroline formulation to (XXXII); similar Δ^1-pyrroline structures exist for other 2-alkyl or 2-aryl-substituted compounds. Even the 2,3-diphenyl substituted pyrroline (XXXIII) exists in

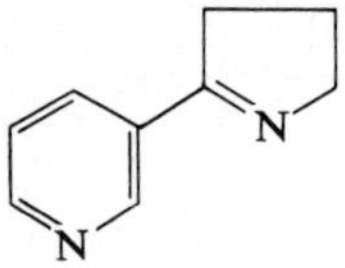

(XXXII) 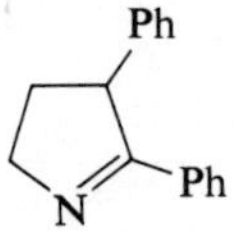 (XXXIII)

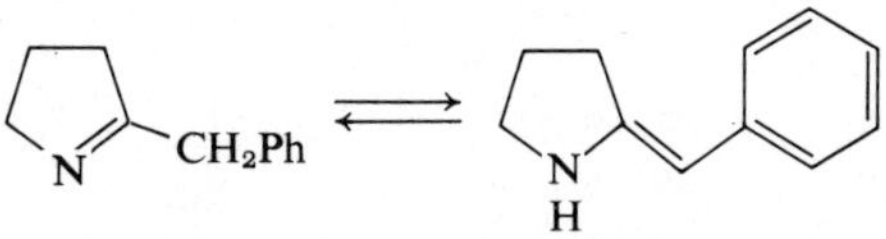

(XXXIV) (XXXV)

the Δ^1-form shown. 2-Benzylpyrroline, however, exists preferentially as the exocyclic enamine (XXXV), rather than the Δ^1-pyrroline form (XXXIV). The majority of 1,2-dialkylpyrrolines exist predominantly in the form containing an exocyclic double bond. However, when the C-2-substituent is a methyl group, the endocyclic Δ^2-form is preferred.

The six-membered ring analogues have not been studied so intently as the pyrrolines, but here again it seems that for simple derivatives the Δ^1-piperideine form is preferred. Thus, the alkaloid γ-coniceine has been shown to exist as the imine tautomer (XXXVI).

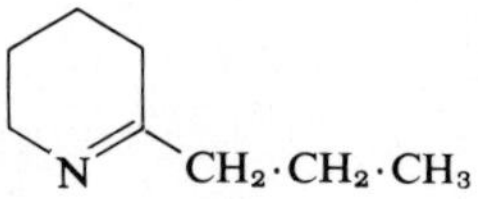

(XXXVI)

3.3.2. Azabicycloalkane enamines. The position of the double bond in enamines derived from 1-azabicycloalkanes seems to be dependent upon several factors. The tautomer (XXXVII) is more stable than (XXXVIII). However, whereas the isomer (XXXIX) is preferred to (XL), the compound (XLI) contains none of its isomer.

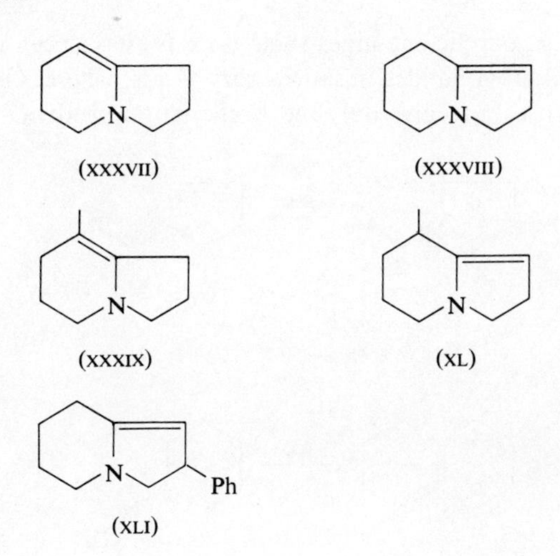

(XXXVII) (XXXVIII)

(XXXIX) (XL)

(XLI)

3.4. Reactions

3.4.1. Electrophilic reagents

Protonation. In the vast majority of the cases that have been studied, protonation of the enamine occurs at C-β to give the iminium ions, e.g.

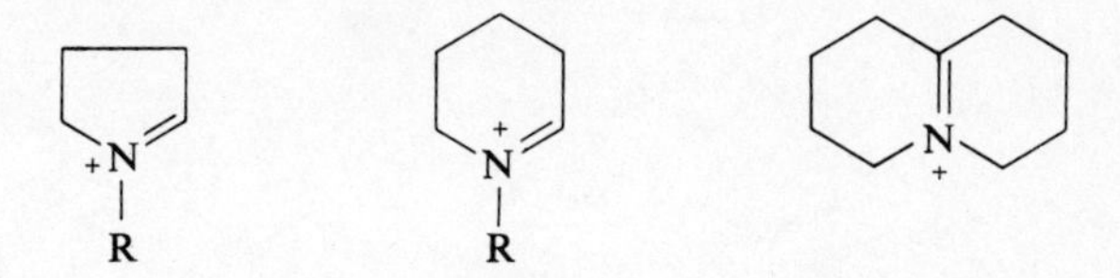

However, some examples of *N*-protonation are known. For example, the Δ^2-piperideine (XLII) yields a mixture of the *C*- and *N*-protonated forms when treated with perchloric acid, whereas quinuclideine (XLIII) gives the *N*-protonated species exclusively; the $>C{=}\overset{+}{N}<$ grouping cannot be generated for steric reasons.

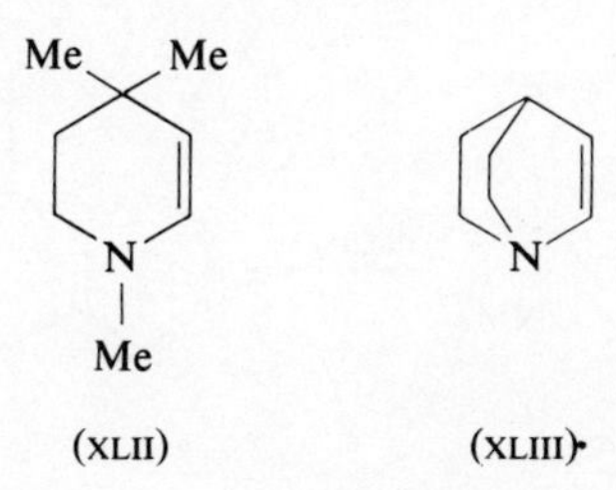

(XLII) (XLIII)

Alkylation. Exocyclic enamines such as (XLIV) have been *C*-alkylated with saturated alkyl halides in satisfactory yields, whereas enamines of the type (XLV) seem to give only the *N*-alkylation product.

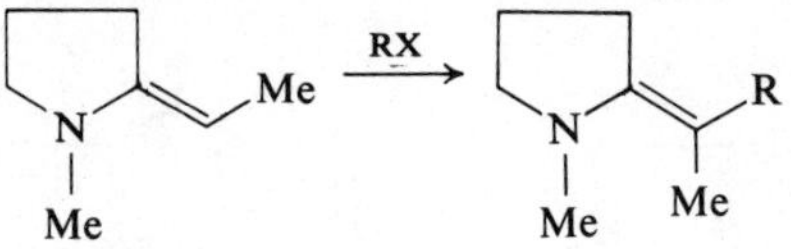

(XLIV)

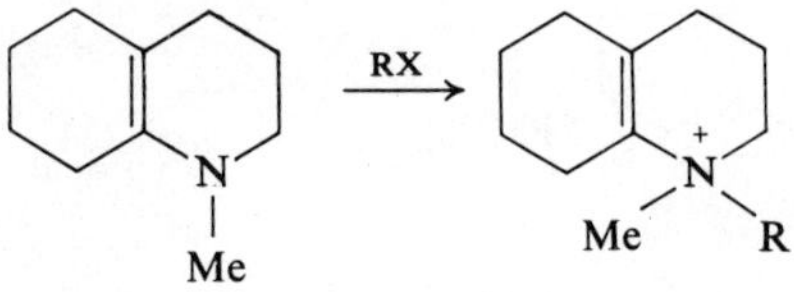

(XLV)

One of the first recorded examples of *C*-alkylation in 1,2-dihydroisoquinolines involved the reaction of dihydroberberine (XLVI) with methyl iodide, but simple 1,2-dihydroisoquinolines (XLVII) have been successfully *C*-alkylated only with benzyl halides.

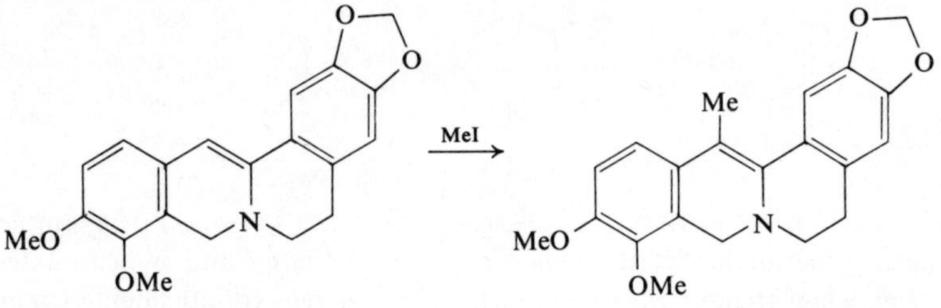

(XLVI)

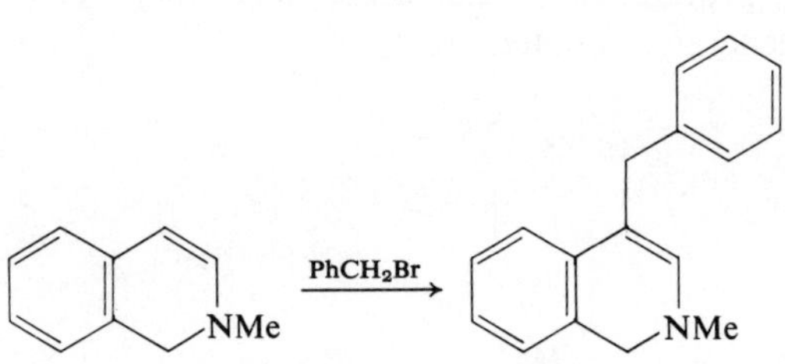

(XLVII)

Azabicycloalkane enamines can give complex results when they are treated with alkyl halides (scheme 3.10), and the *N*-alkylated product,

Scheme 3.10

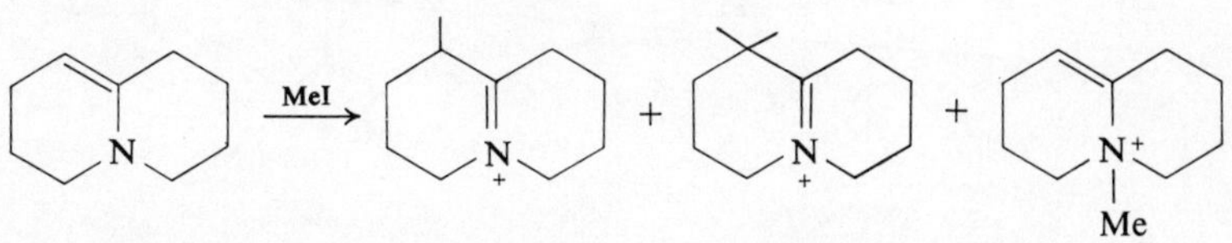

in the case shown, is obtained in 83 per cent yield. Methylation of the enamine (XLVIII) gives the *N*-alkylated material exclusively.

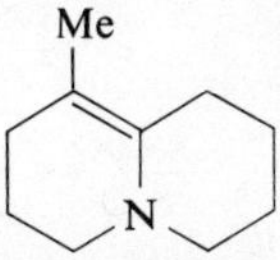

(XLVIII)

When more reactive halides, e.g. allyl halides, are used, higher proportions of *C*-allylation are observed.

Similar results are obtained with indolizidine enamines (scheme 3.11).

Scheme 3.11

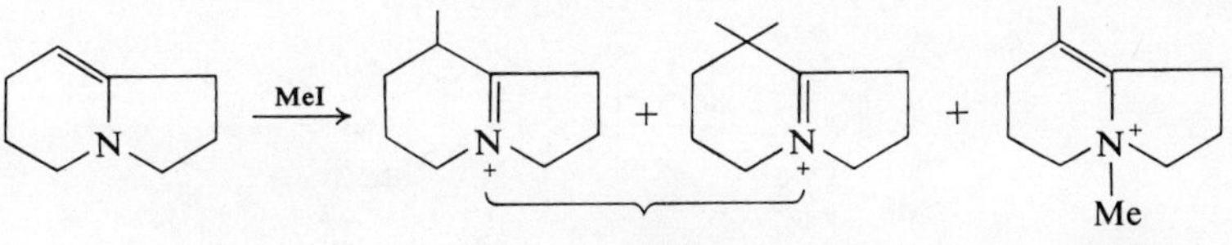

50% ratio of 2:1   50%

Indole is a heterocyclic enamine, and many of its reactions can be interpreted and rationalised in these terms:

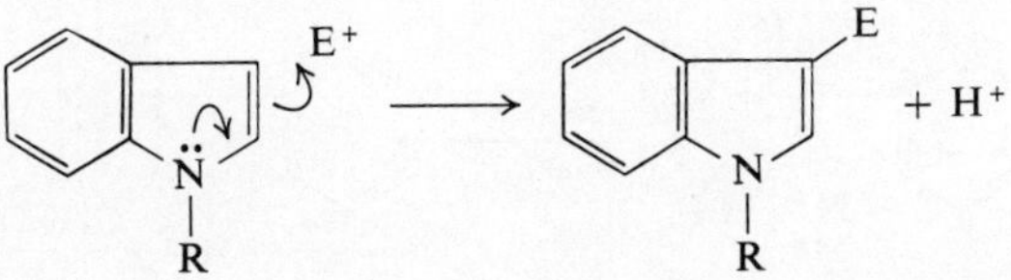

One of the first recorded reactions that can now be interpreted as that of a simple enamine, involves the conversion of (XLIX) into (L).

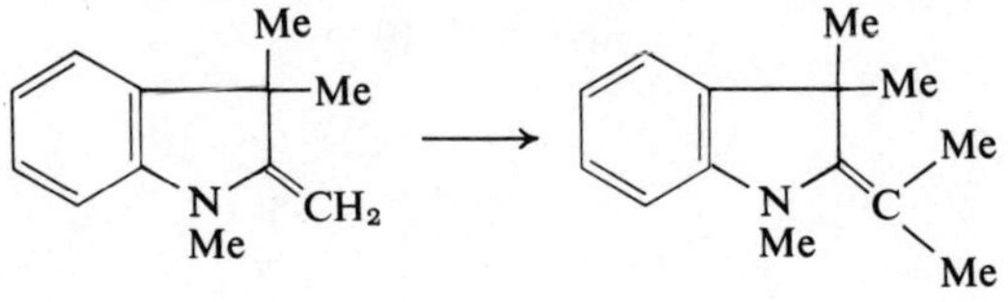

(XLIX) (L)

Acylation. When Δ^1-pyrrolines or Δ^1-piperideines are treated with acyl halides, ring-opening occurs:

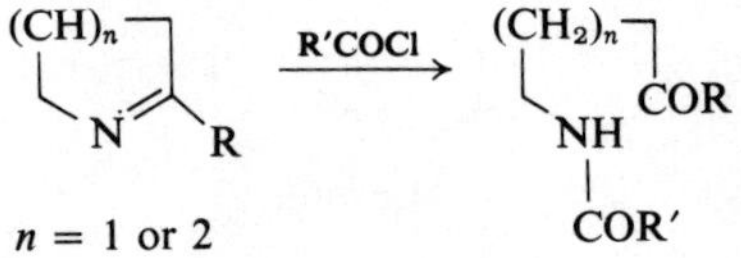

R′COCl

$n = 1$ or 2

Other acylation reactions are summarised in scheme 3.12.

Scheme 3.12

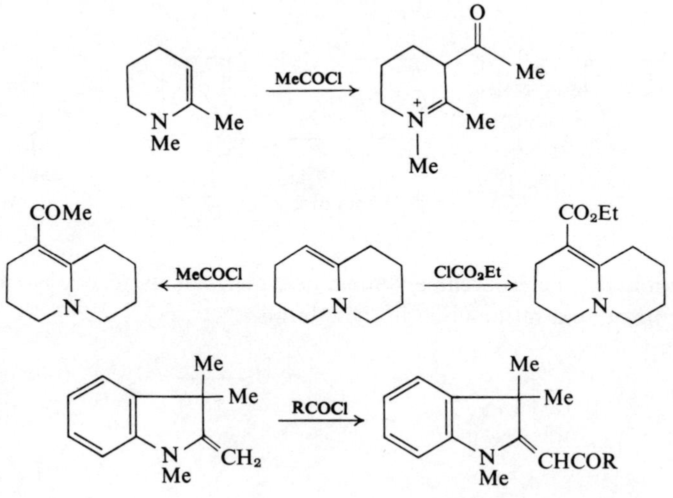

MeCOCl

MeCOCl ClCO$_2$Et

RCOCl

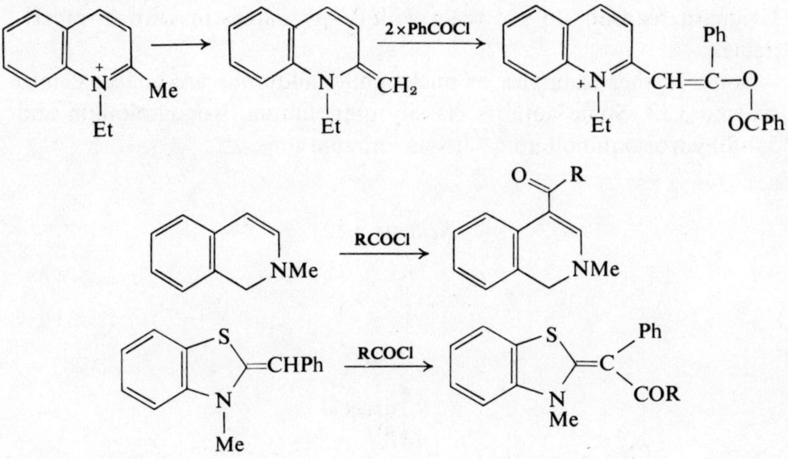

2 × PhCOCl

RCOCl

RCOCl

3.4.2. Nucleophilic reagents. Cyanide ion and Grignard reagents have been added to iminium salts to yield tertiary bases:

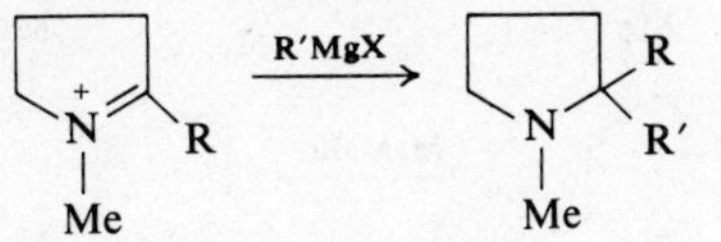

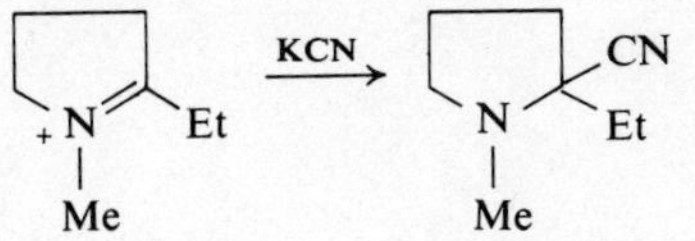

R′MgX

KCN

(LI)

The pseudocyanides such as (LI) decompose on treatment with acids and the iminium ion is regenerated. However, Δ^1-pyrrolines give stable cyano derivatives, where the —C≡N group exhibits its usual reactions:

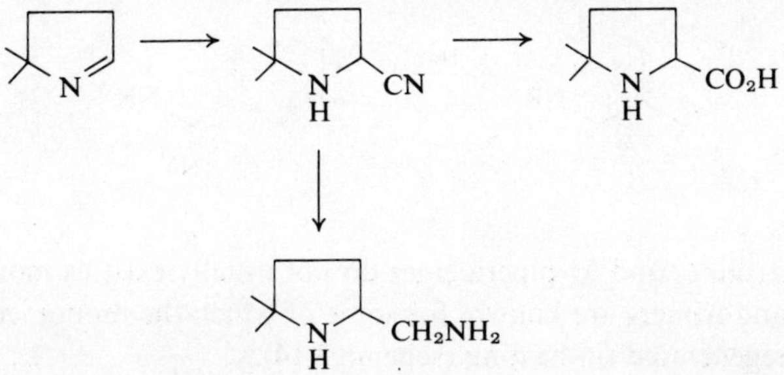

Grignard reagents do not react with Δ^1-pyrrolines or with Δ^1-piperideines.

Some further examples of nucleophilic additions are collected into scheme 3.13. Some authors classify quinolinium, isoquinolinium and 3,4-dihydroisoquinolinium salts as iminium ions.

Scheme 3.13

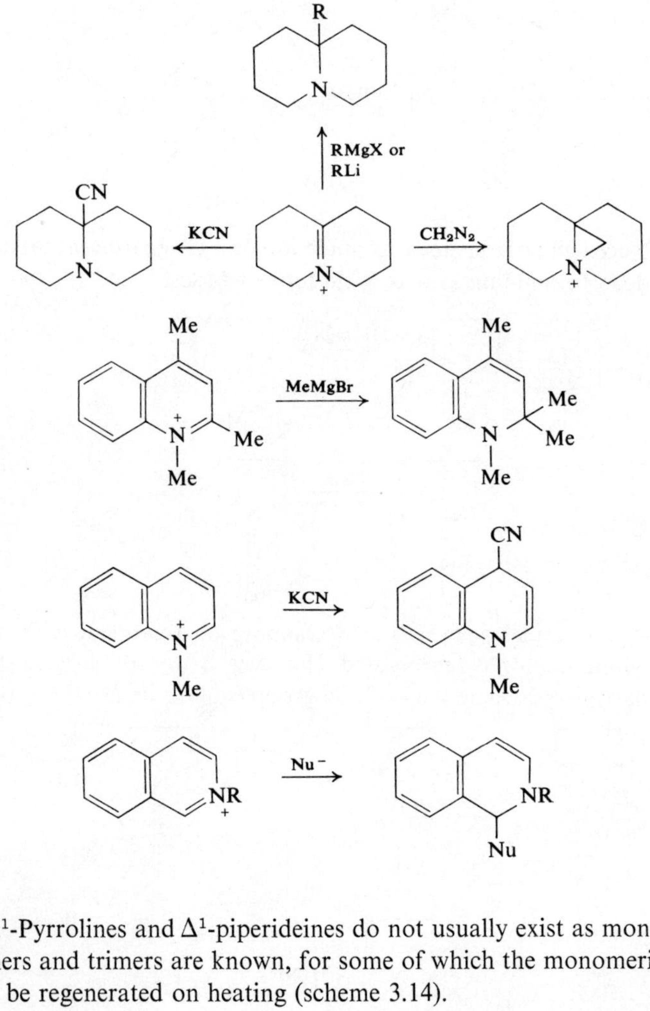

RMgX or RLi

KCN

CH_2N_2

MeMgBr

KCN

Nu^-

Δ^1-Pyrrolines and Δ^1-piperideines do not usually exist as monomers; dimers and trimers are known, for some of which the monomeric form can be regenerated on heating (scheme 3.14).

Scheme 3.14

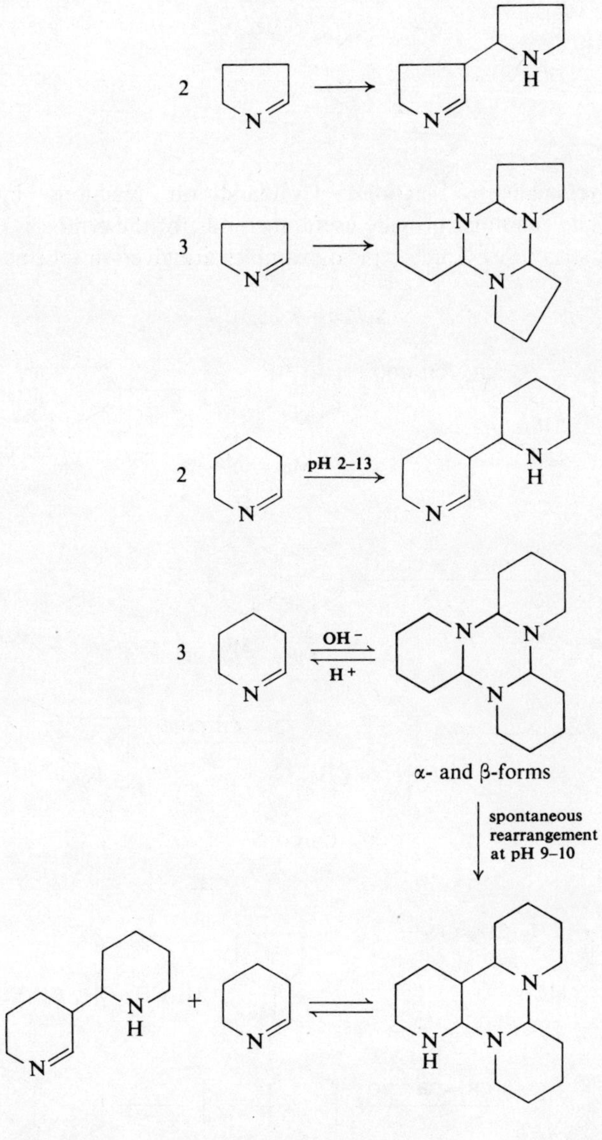

2 →
3 →

2 —pH 2–13→

3 ⇌ (OH⁻ / H⁺)

α- and β-forms

spontaneous rearrangement at pH 9–10

+ ⇌

N-Substituted Δ^2-piperideines also form dimers if C-2 is not substituted; the perchlorate salt (LII) is stable however.

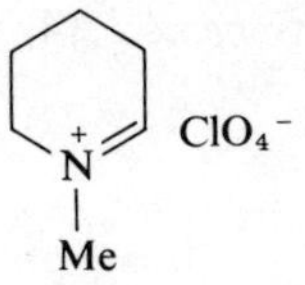

N ClO_4^-

Me

(LII)

3.4.3. Cycloaddition reactions. Cycloaddition reactions involving heterocyclic enamines provide useful methods for the synthesis of more complex structures. Some typical examples are given in scheme 3.15.

Scheme 3.15

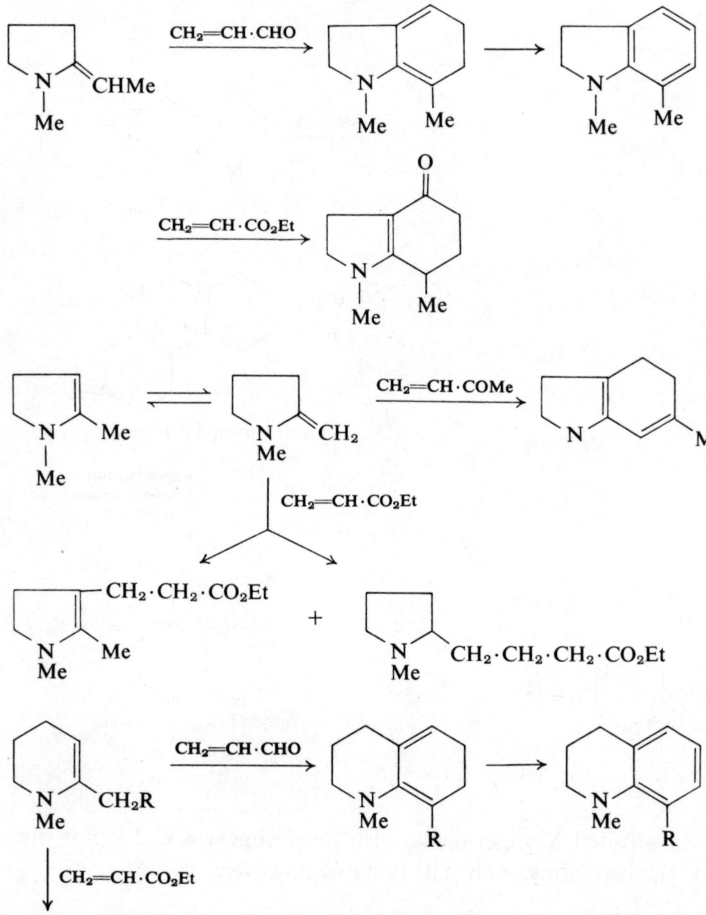

N CHMe Me $\xrightarrow{CH_2{=}CH{\cdot}CHO}$ N Me Me $\longrightarrow$ N Me Me

$\xrightarrow{CH_2{=}CH{\cdot}CO_2Et}$ O N Me Me

N Me Me $\rightleftharpoons$ N CH_2 Me $\xrightarrow{CH_2{=}CH{\cdot}COMe}$ N Me

$CH_2{=}CH{\cdot}CO_2Et$

$CH_2{\cdot}CH_2{\cdot}CO_2Et$ N Me Me + N Me $CH_2{\cdot}CH_2{\cdot}CH_2{\cdot}CO_2Et$

N Me CH_2R $\xrightarrow{CH_2{=}CH{\cdot}CHO}$ N Me R $\longrightarrow$ N Me R

$\downarrow$ $CH_2{=}CH{\cdot}CO_2Et$

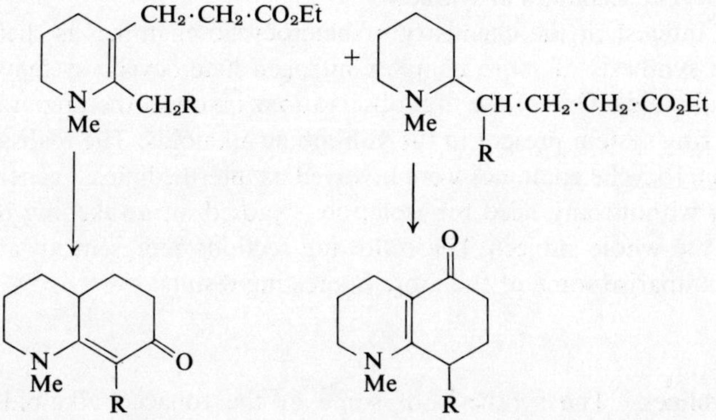

$CH_2 \cdot CH_2 \cdot CO_2Et$ + $CH \cdot CH_2 \cdot CH_2 \cdot CO_2Et$

N Me CH_2R — N Me R

N Me O R — O N Me R

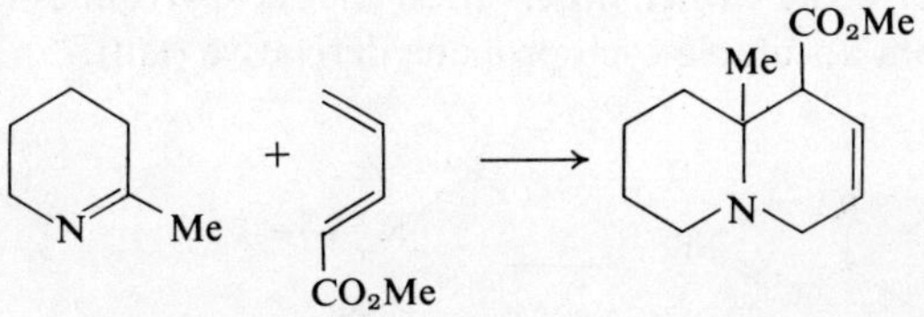

N Me + CO_2Me ⟶ CO_2Me Me N

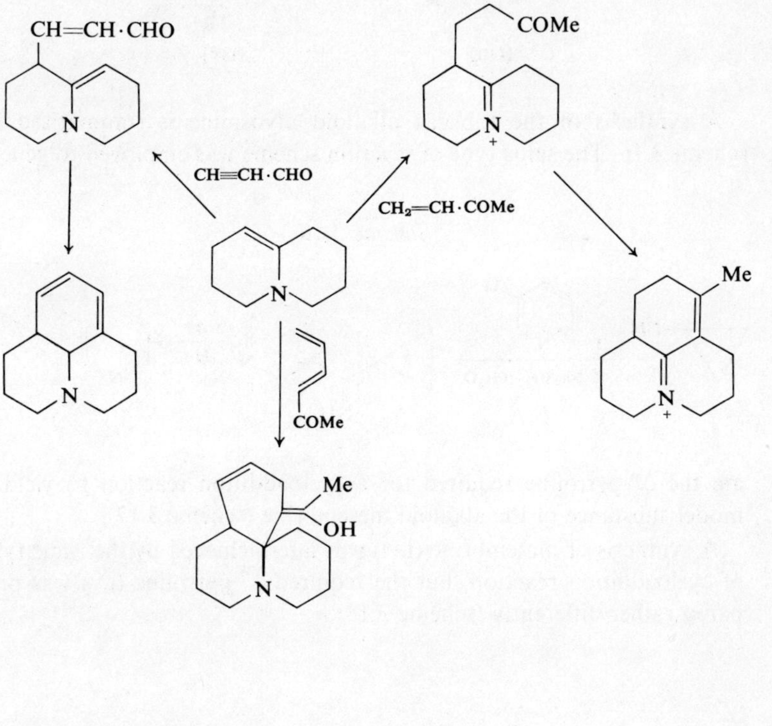

$CH{=}CH \cdot CHO$ — COMe

N — N+

$CH{\equiv}CH \cdot CHO$ — $CH_2{=}CH \cdot COMe$

N — Me

N — N+

COMe

Me OH

N

3.5. Heterocyclic enamines in synthesis

The major interest in the chemistry of heterocyclic enamines is their use for the synthesis of more complex nitrogen heterocyclic systems, especially alkaloids. One of the first observations involved the preparation of the ring system present in the yohimbine alkaloids. The realisation that heterocyclic enamines were involved as intermediates – generated *in situ* without any need for isolation – caused an awakening of interest in the whole subject. The following sections represent an attempt to summarise some of the more interesting results.

3.5.1. Pyrrolines. The synthesis of some of the tobacco alkaloids, and of some of the amaryllidaceae alkaloids (especially mesembrine and mesembrinine) involve the earlier observation that Δ^2-pyrrolines (LIV) can be obtained from a suitable cyclopropane derivative (LIII):

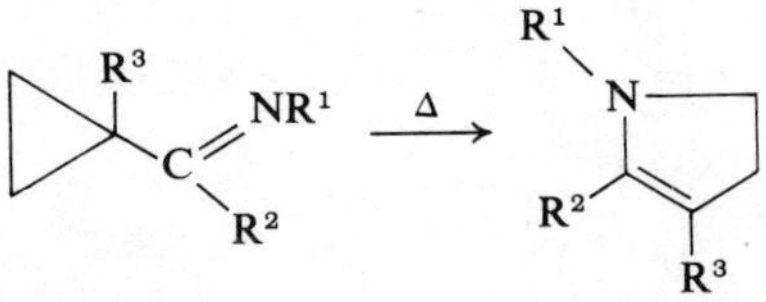

R³ NR¹ C R² Δ → R¹ N R² R³

(LIII) (LIV)

A synthesis of the tobacco alkaloid myosmine is summarised in scheme 3.16. The same type of reaction scheme was employed to gener-

Scheme 3.16

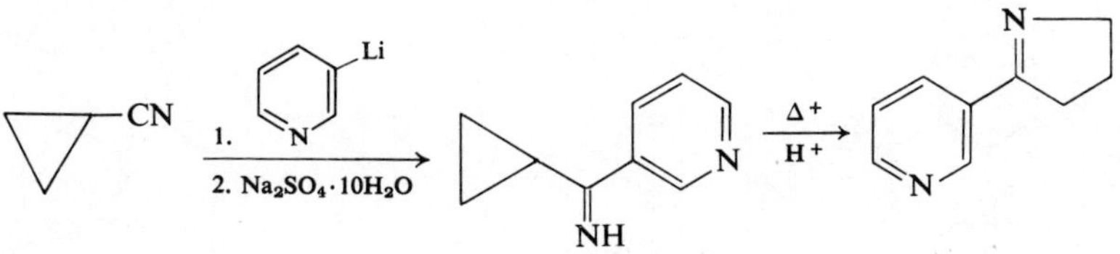

CN 1. Li N 2. $Na_2SO_4 \cdot 10H_2O$ → N NH Δ+ H+ → N N

ate the Δ^2-pyrroline required for a cycloaddition reaction to yield a model substance of the alkaloid mesembrine (scheme 3.17).

A synthesis of mesembrine (LVI) was later achieved by the same type of cycloaddition reaction, but the required Δ^2-pyrroline (LV) was prepared rather differently (scheme 3.18).

Scheme 3.17

Ph
CHO
$MeNH_2$
Ph
H
NMe
Δ
O
N
Me
Me
C=O
$CH{=}CH_2$
Me
N
Ph

Scheme 3.18

O
N
Me
MeO
MeO
Li
OMe
OMe
OH
N
Me
H^+
O
MeO
MeO
N
Me
$CH_2{=}CH \cdot COMe$
in
$CH_3 \cdot CN$
MeO
MeO
N
Me

(LV)

OMe
OMe
O
H
N
Me

(LVI)

Scheme 3.19

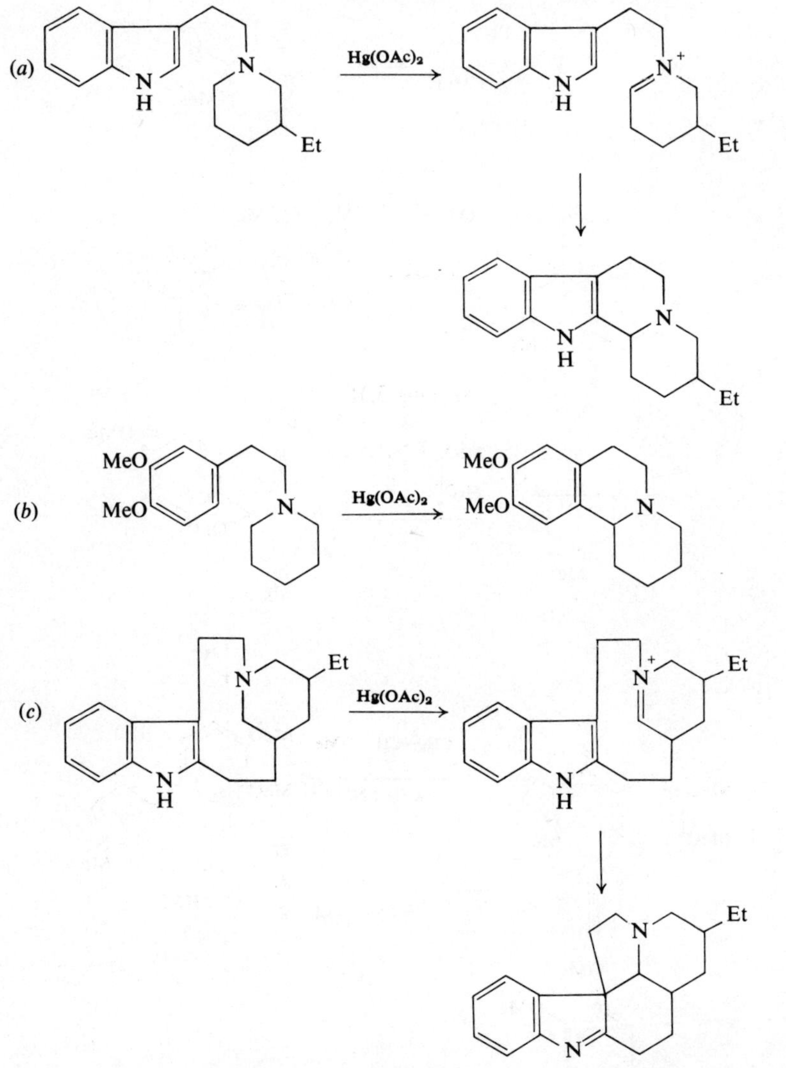

(*a*) Hg(OAc)$_2$ → Et

(*b*) MeO MeO Hg(OAc)$_2$ →

(*c*) Et Hg(OAc)$_2$ →

3.5.2. Pyridine derivatives. Wenkert and his associates have developed some of the enamine chemistry of pyridine derivatives into a very powerful route to complex nitrogen-containing structures. The approach was to generate the iminium ion in the pyridine ring and then to subject it to reactions with nucleophilic reagents at C-α, usually in an intramolecular reaction. The iminium ion was generated by one of the following methods:

(i) reduction of the pyridinium salt with LAH. This proved to be a rather unsatisfactory method.

(ii) oxidation of suitably substituted piperidines with mercuric acetate. Some illustrative examples appear in scheme 3.19. The indole ring itself is a strongly nucleophilic reagent, but the less powerful 3,4-dialkoxybenzene ring has also been successfully used.

(iii) the partial, catalytic reduction of 3-acylpyridinium ions. The acyl group was removed by hydrolysis in a subsequent step in certain examples. The essential concept involved in this method of enamine formation is that in basic solution, a 3-acylpyridinium salt will be converted into a pseudobase which is a vinylogous amide (scheme 3.20). Such a function will be inert to catalytic hydrogenation, so that the overall effect of subjecting a 3-acylpyridinium ion to catalytic hydrogenation in a basic medium will be the *selective* reduction to form a species such as (LVII).

Scheme 3.20

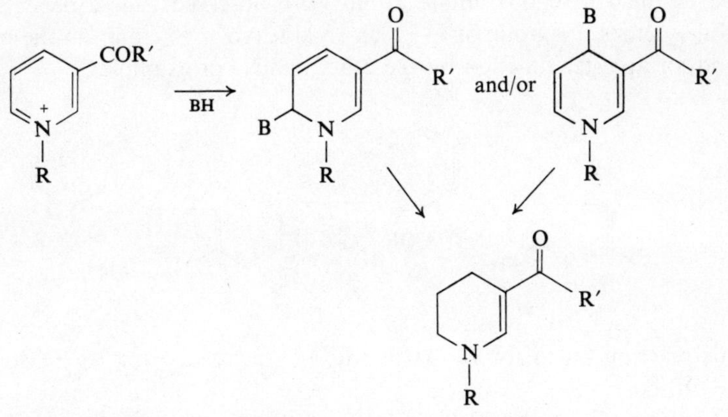

COR′
BH
B N R
O R′ and/or
B O R′
N R
O R′
N R
(LVII)

By an appropriate choice of the functions R and R′, some synthetically useful reactions can be accomplished. The essential steps in a synthesis of the alkaloid eburnanonine (LVIII) are summarised in scheme 3.21.

Scheme 3.21

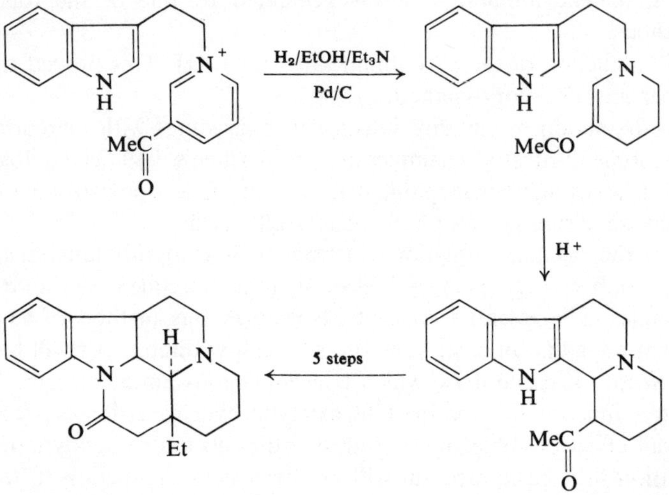

N
H
N^+
MeC
O

H_2/EtOH/Et_3N
Pd/C

N
H
N
MeCO

H^+

H
N
N
O
Et

5 steps

N
H
N
MeC
O

(LVIII)

Some differences in reactivity in the important acid-catalysed cyclisation of the vinylogous amide group were observed. These depended upon whether the group at C-3 was an aldehyde or ketone on the one hand, or an ester function on the other hand. For example:

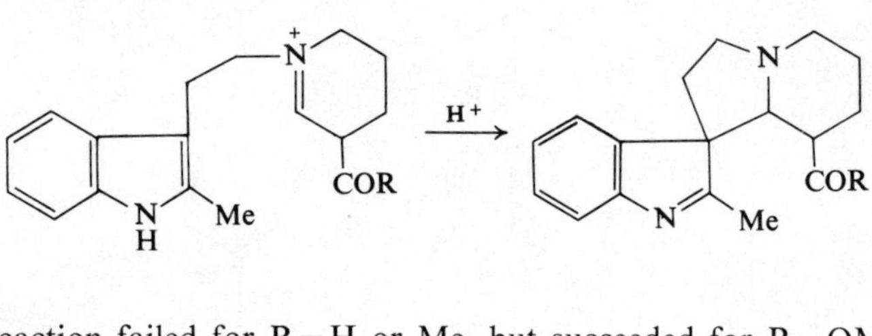

N^+
COR
N
H
Me

H^+

N
COR
N
Me

This reaction failed for R = H or Me, but succeeded for R = OMe or R = OBu^t.

A study of the ultraviolet spectra of these enamines showed that, for those possessing an aldehyde or ketone function at C-3, protonation occurs predominantly on oxygen (LIX) → (LX). However, for a C-3-ester function, protonation occurs at the β-carbon atom with respect to nitrogen, (LXI) → (LXII), thus facilitating the required cyclisation. The problem was also overcome by converting the aldehyde (LIX,

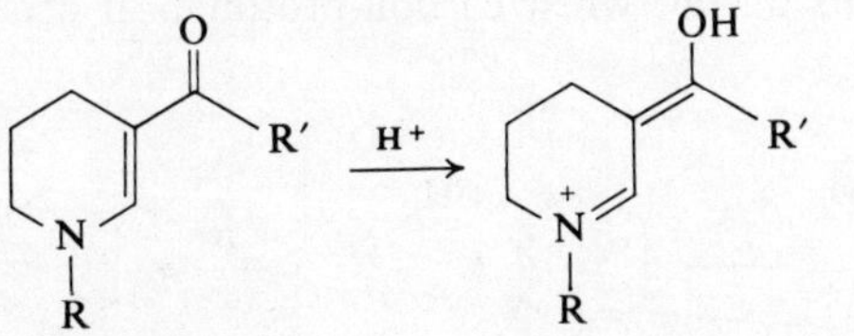

O
R′
N
R
H^+
OH
R′
$\overset{+}{N}$
R

R′ = H or Me

(LIX) (LX)

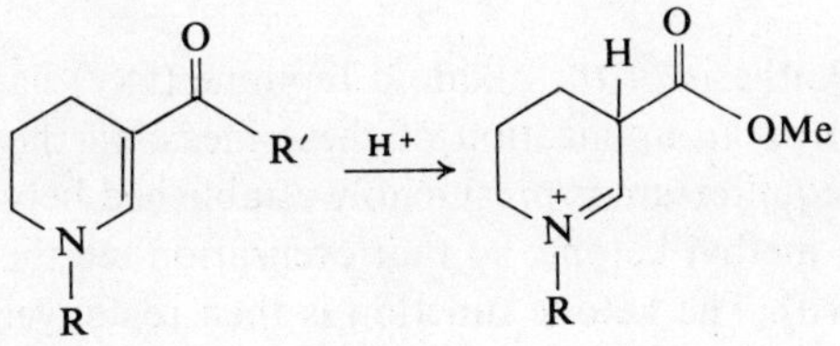

O
R′
N
R
H^+
H
O
OMe
$\overset{+}{N}$
R

R′ = OMe

(LXI) (LXII)

Scheme 3.22

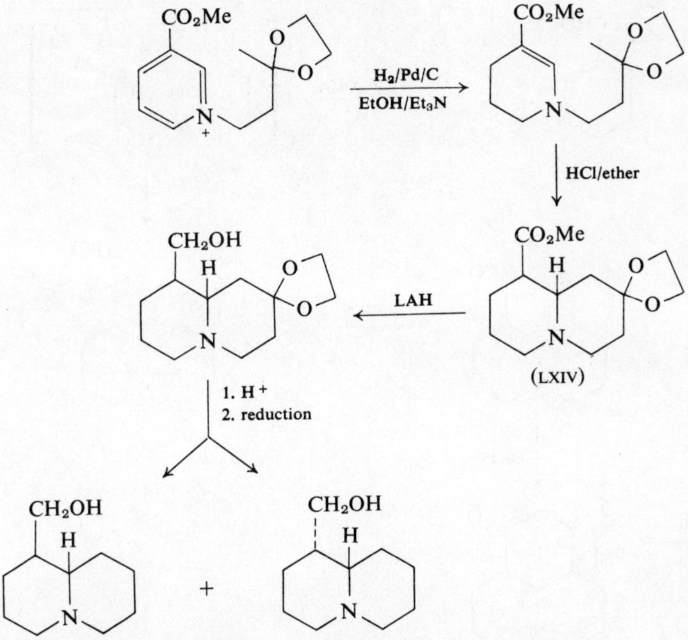

CO_2Me
O
O
$\overset{+}{N}$
H_2/Pd/C
EtOH/Et_3N
CO_2Me
O
O
N

HCl/ether

CH_2OH
H
O
O
N
LAH
CO_2Me
H
O
O
N

(LXIV)

1. H^+
2. reduction

CH_2OH
H
N
+
CH_2OH
H
N

(LXV)

R′ = H) into its acetal, when carbon-protonation occurs to yield the ion (LXIII):

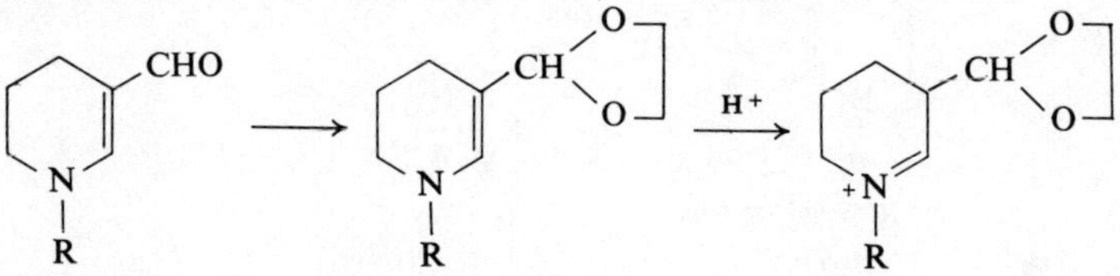

(LXIII)

A simple synthesis of the alkaloid lupinine (LXV) has been achieved (scheme 3.22) by an application of these ideas. In the crucial cyclisation step, an equilibrium is presumably established between the starting ketal and the methyl ketone, so that cyclisation *via* the enol can occur (LXVI) → (LXVII). The ketone function is then re-converted into a ketal (LXIV).

Scheme 3.23

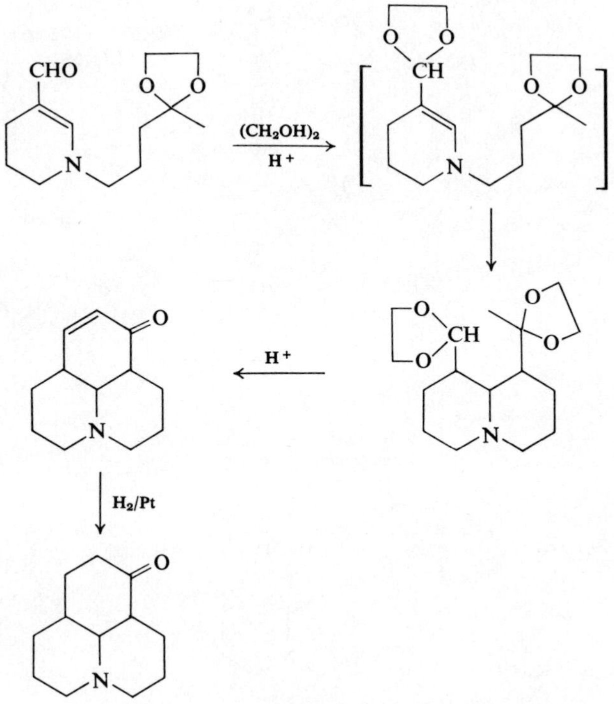

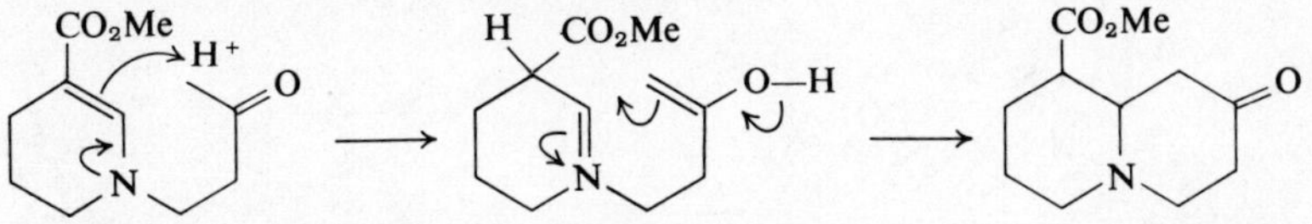

(LXVI) (LXVII)

A useful modification on this theme is illustrated in scheme 3.23.

For some purposes the β-acyl fragment attached to the pyridine ring is not required in the final product, so it can be removed by alkaline hydrolysis, especially if the function is an ester (scheme 3.24).

Scheme 3.24

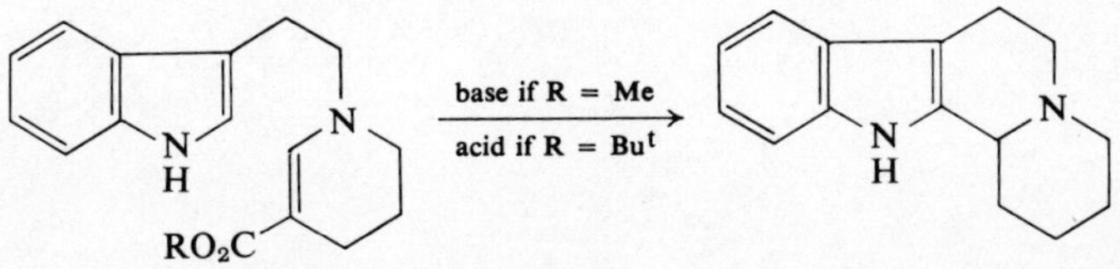

base if R = Me
acid if R = Bu^t

This procedure has been employed by Wenkert in several model studies aimed at synthesis of the corynantheidine alkaloids:

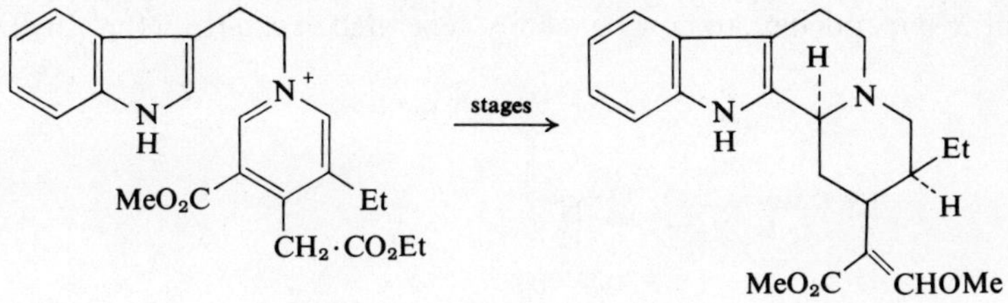

stages

One further illustration of Wenkert's approach is provided by his synthesis of the alkaloid lamprolobine (scheme 3.25).

Scheme 3.25

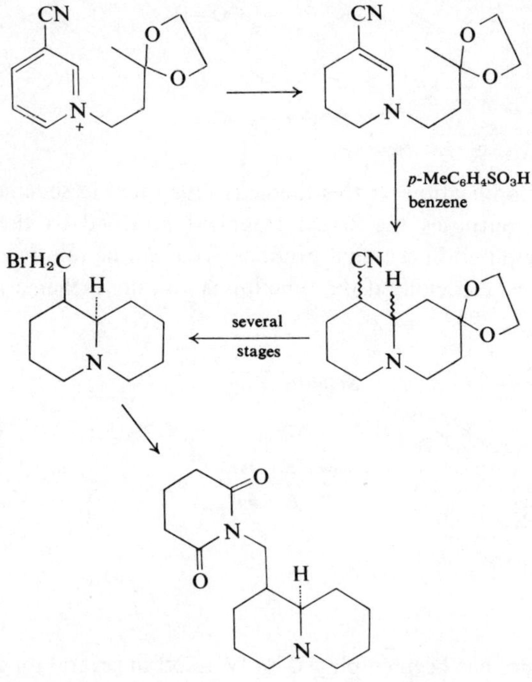

p-$MeC_6H_4SO_3H$
benzene

several stages

(LXVIII)

A side-issue from the above studies on pyridine-derived enamines is that Δ^1-piperideine can now be readily generated *in situ* from the vinylogous amide ester (LXVIII) which is itself easily obtained from nicotinic acid.

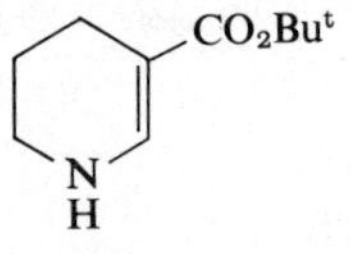

3.5.3. Isoquinoline derivatives. One of the earliest applications of the use of 1,2-dihydroisoquinolines in synthesis, involved the preparation of the carbon–nitrogen skeleton of the yohimbine alkaloids. Several examples of this procedure (scheme 3.26) are now known.

Scheme 3.26

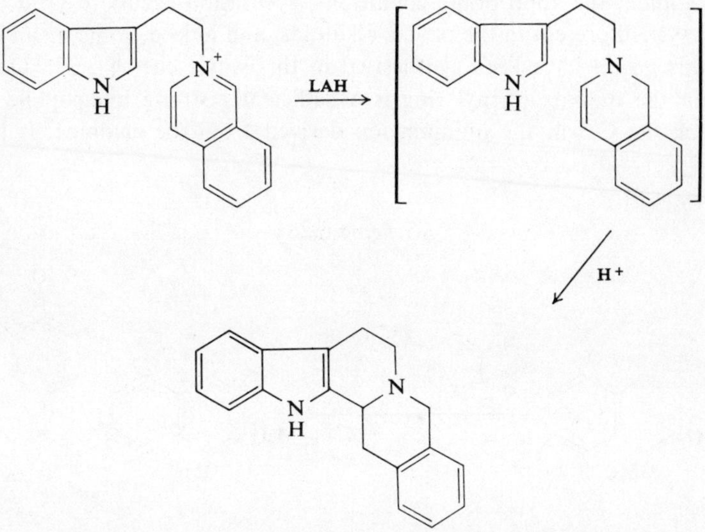

LAH

H^+

Scheme 3.27

MeO MeO $\overset{+}{N}$Me OMe OMe

LAH

MeO MeO NMe OMe OMe

$H_3PO_4/POCl_3$

Δ

OMe OMe MeO MeO NMe

MeO MeO NMe OMe OMe

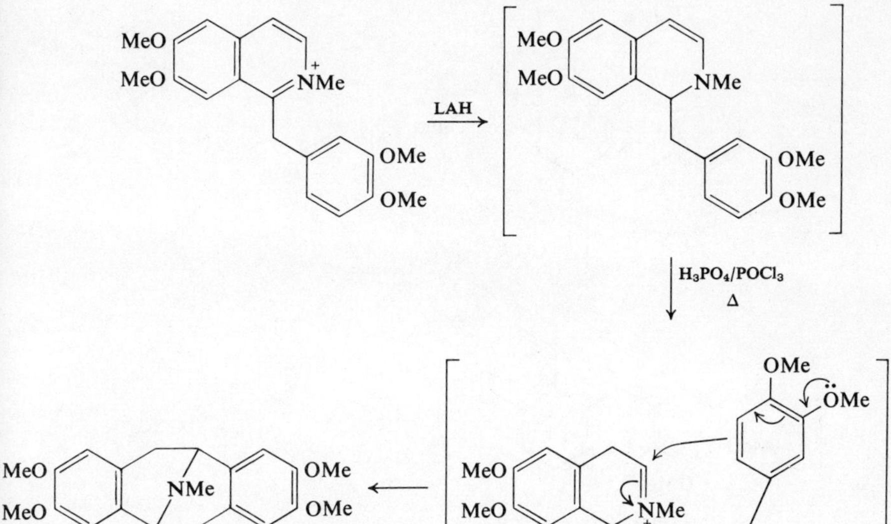

When a 1-benzyl-1,2-dihydroisoquinoline is treated with mineral acids under the appropriate conditions, cyclisation occurs to yield the ring system present in the pavine alkaloids, and indeed, some members of this group have been synthesised in this way (scheme 3.27). Once again the dialkoxyphenyl ring is a sufficiently strong nucleophile to attack the C-3 in the iminium ion derived from the enamine. If the

Scheme 3.28

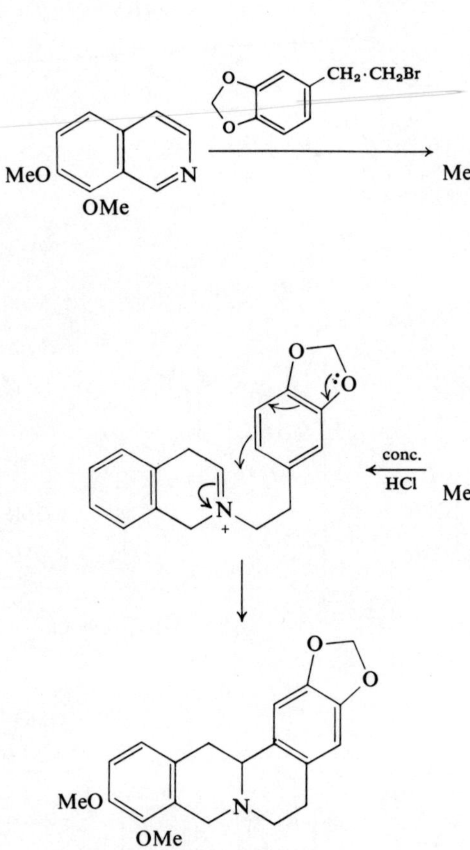

(±)-Tetrahydroberberine

1-benzyl-1,2-dihydroisoquinoline is treated with dilute mineral acid at a lower temperature a different reaction occurs:

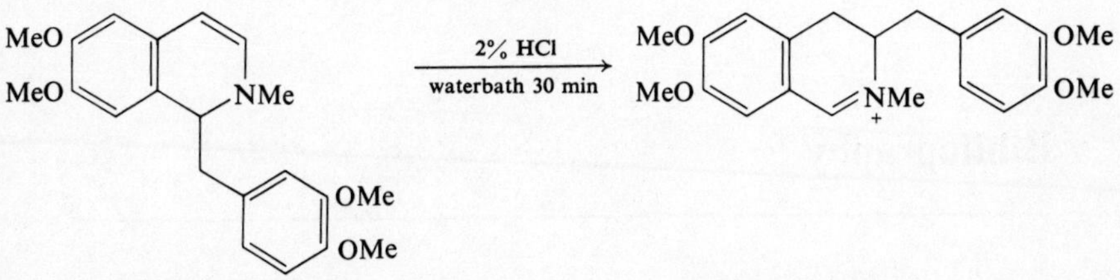

2% HCl
waterbath 30 min

A very effective synthesis of the berberine alkaloids (scheme 3.28) has also been developed based upon the principles outlined in scheme 3.27.

A further extension of this type of reaction to the synthesis of the benzo[c]phenanthridine alkaloids failed:

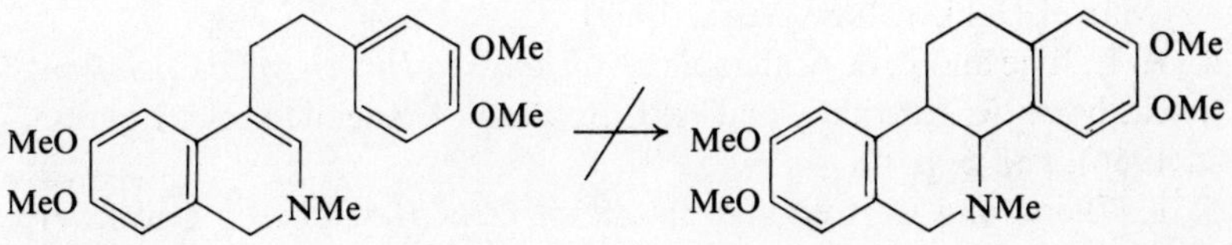

although the closely related ring-closure of (LXIX) to (LXX) was successful.

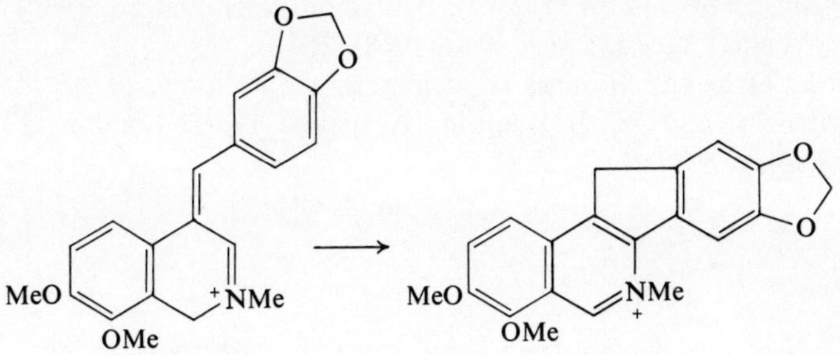

(LXIX) (LXX)

Bibliography

1. G. Stork, A. Brizzolara, H. Landesman, J. Szmuszkovicz and R. Terrell, *J. Am. Chem. Soc.* (1963), **85**, 207.
2. J. Szmuszkovicz in *Advances in Organic Chemistry, Methods and Results*, ed. by R. A. Raphael, E. C. Taylor and H. Wynberg (Interscience, New York, 1963), vol. 4, p. 1.
3. A. G. Cook, ed. *Enamines: Synthesis, Structure and Reactions* (Marcel Dekker, New York, 1969).
4. R. E. Lyle and P. S. Anderson in *Advances in Heterocyclic Chemistry*, ed. by A. R. Katritzky and A. J. Boulton (Academic Press, London, 1966), vol. 6, p. 46.
5. K. Bláha and O. Červinka in *Advances in Heterocyclic Chemistry*, ed. by A. R. Katritzky and A. J. Boulton (Academic Press, London, 1966), vol. 6, p. 147.
6. W. I. Taylor, *Indole Alkaloids: An Introduction to the Enamine Chemistry of Natural Products* (Pergamon Press, Oxford, 1966).
7. E. Wenkert, *Accts. Chem. Res.* (1968), **1**, 78.
8. S. F. Dyke in *Advances in Heterocyclic Chemistry*, ed. by A. R. Katritzky and A. J. Boulton (Academic Press, London, 1972), vol. 14, p. 279.

Index

acetaldehyde enamines, 3
acids, long-chain, synthesis, 27
aldehyde enamines
  C- v. N-alkylation, 15
  reaction with
    alkyl halides, 14
    electrophilic olefines, 31
    nitro-olefines, 39, 40
    α, β-unsaturated aldehydes, 41
    α, β-unsaturated esters, 39
    α, β-unsaturated ketones, 40
    α, β-unsaturated nitriles, 39
alkaloids, synthesis of
  berberine, 88
  corynantheidine, 85
  indole, 80
  lamprolobine, 85
  lupinine, 84
  mesembrine, 78
  myosmine, 78
  pavine, 87
  yohimbine, 86
alkylation
  C- v. N, 17
  of aldehyde enamines, 14
  of dihydroberberine, 70
  of 1, 2-dihydroisoquinolines, 70
  of heterocyclic enamines, 70
  of ketone enamines, 15
  with acyl halides, 22
  with alkyl halides, 14
  with benzyl halides, 17
  with propargyl halides, 17
  with sulphonyl halides, 30
aminal, 3
arylation of enamines, 41

benzyne
  cycloaddition, 42
  reaction with enamines, 42
berberine, 70
berberine alkaloids, synthesis, 88
bromination of enamines, 43

carbenes, reaction with enamines, 50
corynantheidine alkaloids, synthesis, 85
cycloadditions to enamines, 23, 24, 25, 26, 27, 28, 30, 33, 34, 35, 36, 37, 38, 39, 40, 42, 46, 48, 49, 50, 51, 53
cyclobutanes, 34, 39
cyclobutanones, 23
cyclobutenes, 47
cyclohexanone enamines, alkylation of, 11

dehydroindolizidine, 58
dehydroquinolizidines, 56
  acylation, 72
  preparation, 56
Diels–Alder addition to enamines, 49
dihydroberberine, alkylation, 70
1,2-dihydroisoquinolines, 55
  acylation, 73
  alkylation, 70
  preparation, 64
  rearrangement, 89
  use in synthesis, 86
1,4-dihydroquinolines, 55
  preparation, 64
1,3-dipolar additions, 50

enamines,
  acylated, use in synthesis, 27
  acylation, 22
    keten formation, 24
    rearrangement, 23
    with acid halides, 22
    with sulphonyl halides, 30
    with α, β-unsaturated halides, 28, 29
  arylation, 41
  azabicycloalkane
    alkylation, 71
    protonation, 70

enamines (*cont.*)
  azabicycloalkane (*cont.*)
    structure, 68
  basicity, 8, 17
  bromination, 43
  cycloaddition, 46
    Diels–Alder, 49
    1,3-dipolar, 50
    miscellaneous, 51
    with acetylenes, 47
    with acyl halides, 23
    with carbenes, 50
    with diazoalkanes, 53
  dialkylation, 18
  heterocyclic, 55
    acylation, 72
    alkylation, 70
    cycloadditions to, 77
    definition, 55
    preparation, 55
    protonation, 69
    reactions, 69; with cyanide ion, 73; with organometallics, 73
    structure, 67
    uses in synthesis, 78; 1,2-dihydro-isoquinolines, 86; pyrrolines, 78; piperideines, 80
  hydrolysis, 8
  indolizidine, alkylation, 71
  infrared spectra, 10
  cis isomer, preparation, 7
  geometrical isomerism, 7
  hydroboration of, 45
  nuclear magnetic resonance spectra, 10, 11
  protonation, 8
  preparation, 1
    from aldehydes and ketones, 1
    from imines, 5
    from iminium salts, 6
    from ketals, 5
  reactions, 14
  reaction with
    acetylenes, 47
    acyl halides, 22
    aldehydes, 42, 43
    alkyl halides, 14
    aryl halides, 41
    benzyne, 42
    carbenes, 50
    cyanide ions, 54
    cyanogen halides, 44
    diazoalkanes, 52, 53
    diazonium salts, 44
    electrophilic olefines, 31
    electrophilic reagents, 14
    halogens, 43
    hydride ions, 52
    lithium aluminium hydride, 52
    miscellaneous reagents, 41
    nucleophiles, 51
    organometallic compounds, 52
    sodium borohydride, 52
  reactivity, 11
  resonance, 7
  structure, 10
  sulphonation, 30
  ultraviolet spectra, 9
  Vilsmeier reaction, 28

indoles, 55
  alkylation, 71
indole alkaloids, synthesis, 80
indolizidine, 58
imines, 1, 5
  reaction with
    Meerwein's salt, 5
    Grignard reagent, 6

ketone enamines, reaction with
  alkyl halides, 15, 16
  diketen, 24
  electrophilic olefines, 31
  keten, 24
  nitro-olefines, 38
  α, β-unsaturated aldehydes, 36
  α, β-unsaturated esters, 32
  α, β-unsaturated ketones, 34
  α, β-unsaturated nitriles, 32

lactams, reaction with organometallic compounds, 65
lithium aluminium hydride, reduction of
  isoquinolines, 64
  pyridines, 61
  pyrroles, 65
  quinolines, 64
lamprolobine, synthesis, 85
lupinine, synthesis, 84

mercuric acetate oxidation
  mechanism of, 57
  preparation of enamines by, 56
  stereochemistry of, 57
  of piperidines, 58
  of pyrrolidines, 60

mercuric acetate oxidation (*cont.*)
of tertiary amines, 56
mesembrine, synthesis, 78
2-methylcyclohexanone, enamine, 11
3-methylcyclohexanone, enamine, 12
myosmine, synthesis, 78

nitro-olefins, reaction with
aldehyde enamines, 39, 40
ketone enamines, 38

olefines, electrophilic
cycloaddition, 76
reaction with aldehyde enamines, 31
reaction with ketone enamines, 31
α, β-olefinic aldehydes
Diels–Alder addition, 49
reaction with aldehyde enamines, 41
reaction with ketone enamines, 36
α, β-olefinic esters, reaction with
aldehyde enamines, 36, 39
ketone enamines, 32
α, β-olefinic ketones, reaction with
aldehyde enamines, 40
ketone enamines, 34
α, β-olefinic nitriles, reaction with
aldehyde enamines, 39
ketone enamines, 32
oxidation, with mercuric acetate, 56

pavine alkaloids, synthesis, 87
Δ^1-piperideines, 68
acylation, 72
dimers, 75
trimers, 75
Δ^2-piperideines, 55
acylation, 72
cycloaddition to, 76
dimers of, 75
preparation of
from lactams, 65
by reduction, 61
protonation, 69
use in synthesis, 80
piperidines, oxidation with mercuric acetate, 59
pyridines, partial reduction of, 61
pyrrolidines, oxidation with mercuric acetate, 60
Δ^1-pyrrolines, 67
acylation, 72
dimers, 74
trimers, 74
reaction with cyanide ion, 73
reaction with Grignard reagents, 74
Δ^2-pyrrolines, 55
preparation from lactams, 65
structure, 68
use in synthesis, 78

quinolizidine, 56
quinuclideine, protonation, 69

reduction
isoquinolines, 64
partial, 61
pyridines, 61
pyrroles, 65
quinolines, 64

sparteine, 58
sodium borohydride reduction of
pyridines, 61
isoquinolines, 64
α-tetralone enamines, 4
alkylation, 20
β-tetralone enamines, 18

Vilsmeier reaction, 28

yohimbine alkaloids, synthesis, 86